# Polymers and Functional Materials for a Cleaner Environment and Human Health

*Sustainable Technologies and New Research*

# Polymers and Functional Materials for a Cleaner Environment and Human Health

*Sustainable Technologies and New Research*


**Ruby Varghese, PhD**

**Ann Rose Abraham, PhD**

**Shrikaant Kulkarni, PhD**

**A. K. Haghi, PhD**

*Editors*



AAP | APPLE ACADEMIC PRESS

First edition published 2025

**Apple Academic Press Inc.**
1265 Goldenrod Circle, NE,
Palm Bay, FL 32905 USA

760 Laurentian Drive, Unit 19,
Burlington, ON L7N 0A4, CANADA

**CRC Press**
2385 NW Executive Center Drive,
Suite 320, Boca Raton FL 33431

4 Park Square, Milton Park,
Abingdon, Oxon, OX14 4RN UK



*Apple Academic Press exclusively co-publishes with CRC Press, an imprint of Taylor & Francis Group, LLC*



**Library and Archives Canada Cataloguing in Publication**

CIP data on file with Canada Library and Archives

**Library of Congress Cataloging-in-Publication Data**

CIP data on file with US Library of Congress

ISBN: 978-1-77964-365-0 (hbk)
ISBN: 978-1-77952-029-6 (pbk)
ISBN: 978-1-00363-960-2 (ebk)

# About the Editors

**Ruby Varghese, PhD**
*Assistant Professor, Department of Chemistry and Biochemistry, School of Sciences, Jain Deemed to be University, Bengaluru, Karnataka, India*

Ruby Varghese, PhD, is currently working as an Assistant Professor of Biochemistry (Department of Chemistry and Biochemistry) at the School of Sciences, Jain Deemed to be University, Bengaluru, Karnataka, India. She has several publications to her credit in many peer-reviewed high-impact journals of international repute, such as *Carbohydrate Polymers, Journal of Polymer Research, Current Drug Targets, Materials Science and Engineering: C, Colloids and Surfaces B: Biointerfaces, Results in Materials, Journal of Clinical Medicine, 3 Biotech, ACS Biomaterials Science & Engineering, Macromolecular Bioscience,* and *International Journal of Medicinal Mushrooms*. She is a reviewer for national and international journals and has authored several book chapters. She has research experience of seven years and teaching experience of eight years. She has collaborated with various national and international laboratories, such as the Central Institute of Fisheries Technology, Kochi, India; the National Research Centre for Grapes, Maharashtra, India; and the United Arab Emirates University. Her intellectual capability has been applauded, and her work earned her awards from the Kerala Science Congress, Swadeshi Science Congress, and other national universities. Dr. Varghese received her MSc, and PhD degrees in Biochemistry from the Mar Athanasios College for Advanced Studies Tiruvalla Mahatma Gandhi University, Kerala, India. She has expertise in the fields of tissue engineering, cancer biology, wound healing and diabetes, and, biopolymeric nanocomposites, etc.

**Ann Rose Abraham, PhD**

*Assistant Professor, Sacred Heart College (Autonomous), Thevara, Kochi, Kerala, India*

Ann Rose Abraham, PhD, is currently an Assistant Professor at the Department of Physics, Sacred Heart College (Autonomous), Thevara, Kochi, Kerala, India. She has expertise in the field of condensed matter physics, nanomagnetism, multiferroics and polymeric nanocomposites, etc. She has research experience at various reputed national institutes, including Bose Institute, Kolkata, India, SAHA Institute of Nuclear Physics, Kolkata, India,and UGC-DAE CSR Centre, Kolkata, India. She has collaborated with various international laboratories. She is a recipient of a young researcher award in physics and best paper awards—2020 and 2021. She has served as an assistant professor and examiner at the Department of Basic Sciences, Amal Jyothi College of Engineering, APJ Abdul Kalam Technological University, Kerala, India. Dr. Abraham is a frequent speaker at national and international conferences. She has a good number of publications to her credit in many peer-reviewed high-impact journals of international repute. She has authored many book chapters and edited more than 10 books with Taylor and Francis, Elsevier, etc.

**Shrikaant Kulkarni, PhD**

*Adjunct Professor, Faculty of Business, Victorian Institute of Technology, Melbourne, Australia; Adjunct Professor, Centre of Research Outcome and Impact, Chitkara University, Punjab, India*

Shrikaant Kulkarni, PhD, is currently an Adjunct Professor in the Faculty of Business at the Victorian Institute of Technology, Melbourne, Australia, as well as an Adjunct Professor at the Centre of Research Outcome and Impact, Chitkara University, Punjab, India. Dr. Kulkarni has been a senior academician and researcher for over four decades. He has delivered invited lectures and conducted sessions at national and international conferences and faculty development programs. He has guided many major and minor projects in engineering chemistry, green chemistry, nanotechnology, analytical chemistry, catalysis, chemical engineering materials, industrial organization, management, and AI. He has published over 100 research papers in national and international journals and conferences of repute. He has authored more than 50 book chapters. He has edited 25 books published by international publishers. Another ten books are in the offing. He authored four textbooks in engineering chemistry. Dr. Kulkarni possesses MSc, MPhil, and PhD degrees in Chemistry apart from master's degrees in Economics, Business Management, and Political Science. He has expertise in materials science, green chemistry and engineering, analytical chemistry, green nanoscience, nanotechnology, AI, and computational intelligence. He has expertise in wastewater treatment, green and analytical chemistry, and advanced areas in chemical engineering and materials science. He published two patents and applied for two more.

**A. K. Haghi, PhD**
*Research Associate, Department of Chemistry, University of Coimbra, Portugal*

A. K. Haghi, PhD, is a retired professor and has written, co-written, edited, or co-edited more than 1000 publications, including books, book chapters, and papers in refereed journals with over 3800 citations and an h-index of 32, according to the Google Scholar database. He is currently a research associate at the University of Coimbra, Portugal. Professor Haghi has received several grants, consulted for several major corporations, and is a frequent speaker to national and international audiences. He is the Founder and former Editor-in-Chief of the *International Journal of Chemoinformatics and Chemical Engineering* and *Polymers Research Journal.* Professor Haghi has acted as an editorial board member of many international journals. He has served as a member of the Canadian Research and Development Center of Sciences & Cultures. He has supervised several PhD and MSc theses at the University of Guilan (UG) and co-supervised international doctoral projects. Professor Haghi holds a BSc degree in urban and environmental engineering from the University of North Carolina (USA) and holds two MSc, one in mechanical engineering from North Carolina State University (USA) and another one in applied mechanics, acoustics, and materials from the Université de Technologie de Compiègne (France). He was awarded a PhD in engineering sciences at Université de *Franche-Comté* (France). He is a regular reviewer for leading international journals.

# Contents

# Contributors

**Jiji Abraham**
Vimala College (Autonomous), Thrissur, Kerala, India

**Jijo Abraham**
St. Aloysius College, Elthuruth, Thrissur, India

**Easwar Vivek Anand**
Department of Physics, Catholicate College, Pathanamthitta, Kerala, India

**Srilakshmi P. Bhasker**
Vimala College (Autonomous), Thrissur, Kerala, India

**Christian I. Cano-Gómez**
Facultad de Estudios Profesionales Zona Huasteca, Universidad Autónoma de San Luis Potosí. Ciudad Valles, San Luis Potosí, México

**Saliya C. S.**
Vimala College (Autonomous), Thrissur, Kerala, India

**Jincemon Cyriac**
Marian College Kuttikanam Autonomous, Kuttikanam, Peermade, Idukki, Kerala, India

**Agampodi Sandali Inupama De Zoysa**
Department of Chemistry & Biochemistry, Jain Deemed to be University, Bengaluru, India

**Dhilna Francis**
Department of Zoology, Sacred Heart College (Autonomous), Cochin, Kerala, India

**Bhaskar Gaonkar**
Department of Chemistry & Biochemistry, Jain Deemed to be University, Bengaluru, India

**Sandy R. García-Rodríguez**
Facultad de Estudios Profesionales Zona Huasteca, Universidad Autónoma de San Luis Potosí. Ciudad Valles, San Luis Potosí, México

**Neenu George**
Department of Chemistry, St. Joseph's College, Moolamattom, Idukki, Kerala, India

**Priyanka Ghow**
Department of Chemistry and Biochemistry, School of Sciences, Jain University, Bangalore, India

**Sneha James**
School of Nanoscience and Nanotechnology, Mahatma Gandhi University, Kottayam, India

**Cintil Jose**
Department of Chemistry, Newman College, Thodupuzha, Idukki, Kerala, India

**Jithin Joy**
Department of Chemistry, Newman College, Thodupuzha, Idukki, Kerala, India

**Nimitha K. C.**
Vimala College (Autonomous), Thrissur, Kerala, India
Christ College (Autonomous), Irinjalakuda, Kerala, India

**Joshy K. S.**
School of Nanoscience and Nanotechnology, Mahatma Gandhi University, Kottayam, India

**Sreelakshmi K. S.**
School of Nanoscience and Nanotechnology, Mahatma Gandhi University, Kottayam, India

**Keerthana M.**
Department of Chemistry and Biochemistry, School of Sciences, Jain University, Bangalore, India

**Soosen Samuel M.**
Department of Physics, Catholicate College, Pathanamthitta, Kerala, India

**Jobi J. Malamel**
Department of Zoology, Sacred Heart College (Autonomous), Cochin, Kerala, India

**Alain Martínez-Pérez**
Departamento de Ingeniería Química y Bioquímica, TecNM/Instituto Tecnológico de Durango. Durango, Durango, México

**Sunny Mathew**
Marian College Kuttikanam Autonomous, Kuttikanam, Peermade, Idukki, Kerala, India

**Rekha Mirle**
Department of Chemistry and Biochemistry, School of Sciences, Jain University, Bangalore, India

**Gayathri M. J.**
Marian College Kuttikanam Autonomous, Kuttikanam, Peermade, Idukki, Kerala, India

**Sadasivan M. J.**
Marian College Kuttikanam Autonomous, Kuttikanam, Peermade, Idukki, Kerala, India

**Diana B. Muñiz-Márquez**
Facultad de Estudios Profesionales Zona Huasteca, Universidad Autónoma de San Luis Potosí. Ciudad Valles, San Luis Potosí, México

**Parvathy M. V.**
School of Nanoscience and Nanotechnology, Mahatma Gandhi University, Kottayam, India

**Yashwanth Narayana**
Department of Chemistry & Biochemistry, Jain Deemed to be University, Bengaluru, India

**Shashanka. K. Prasad**
Department of Biotechnology and Bioinformatics, JSSAHER, Mysuru, Karnataka, India

**Roopashree R.**
Department of Chemistry and Biochemistry, School of Sciences, Jain University, JC Road, Bangalore, India

**Nivya Roy**
Carmel College (Autonomous), Mala, Thrissur, India

**Susan Samuel**
Vimala College (Autonomous), Thrissur, Kerala, India

**Aiswarya Sathian**
College of Science and Engineering, James Cook University, Townsville, Queensland, Australia
International and Inter University Centre for Nanoscience and Nanotechnology, Mahatma Gandhi University, Kottayam, India

**Pooja Shaji**
Department of Physics, Catholicate College, Pathanamthitta, Kerala, India

**Eunica K. Sunny**
Vimala College (Autonomous), Thrissur, Kerala, India

**Unnikrishnan T. G.**
Department of Mechanical Engineering, Viswajyothi College of Engineering and Technology, Kerala, India

**Sabu Thomas**
School of Nanoscience and Nanotechnology, Mahatma Gandhi University, Kottayam, India
International and Inter University Centre for Nanoscience and Nanotechnology, Mahatma Gandhi University, Kottayam, India

**Swarupa V.**
Affiliated to Department of Chemistry and Biochemistry, School of Sciences, Jain University, JC Road, Bangalore, India

**Namitha Vijay**
Department of Biotechnology and Bioinformatics, JSSAHER, Mysuru, Karnataka, India

**Sheeja V. N.**
Marian College Kuttikanam Autonomous, Kuttikanam, Peermade, Idukki, Kerala, India

**Jorge E. Wong Paz**
Facultad de Estudios Profesionales Zona Huasteca, Universidad Autónoma de San Luis Potosí. Ciudad Valles, San Luis Potosí, México

# Preface

This edited volume offers a comprehensive exploration of integrating polymers and functional materials for a cleaner environment and human health using different technologies and new research, catering to a diverse audience ranging from students and budding engineers to seasoned professionals in the industrial sectors. The aim of this volume is to cover multidisciplinary sustainability topics from the perspective of sustainability issues in materials science, health, and environmental crises. This unique research-oriented book is a vital source for polymers postgraduate students and chemical and environmental engineers. It places a strong emphasis on the relationship between engineering and product quality, a sustainable future, and safety development. It explores case studies related to our growing demand from the perspectives of disciplines ranging from environmental science to public health.

This volume links theory and practice and provides comprehensive, up-to-date technical information on sustainable technologies. The chapters cover a wide array of crucial topics, each meticulously crafted to provide valuable insights into the complex sustainable development methods. It addresses the paramount concern of safety in the polymers and functional materials, tackling challenges and opportunities in ensuring the quality and integrity of such products. This research-oriented book also examines the key issue of sustainable environment maintaining safety and health as it becomes more varied and complex. The book also looks at sustainable materials technologies and how they are combined to achieve the right balance of safety and life quality.

This book considers topics in light of factors such as health and environmental issues and serves as an invaluable resource for students, educators, researchers, and professionals engaged in the sustainable future goals.

Whether you are an aspiring engineer, a seasoned expert, or a curious consumer, this comprehensive guide offers a deep understanding of the sustainable development complexities and innovations, addressing critical challenges and charting a course toward a safer and sustainable future.

# CHAPTER 1

# Sustainable Polymer Composites: Classification and Its Applications

SHEEJA V. N., SADASIVAN M. J., GAYATHRI M. J., SUNNY MATHEW, and JINCEMON CYRIAC

*Marian College Kuttikanam Autonomous, Kuttikanam, Peermade, Idukki, Kerala, India*

## ABSTRACT

Polymer matrix composites (PMCs) are a family of materials that are made up of polymer matrix and reinforcement. Polymer composites have a wide range of applications including aviation, automotive, and construction. This chapter begins with an overview of what a PMC is and classifies them based on their polymer matrix constituents including metal matrix, ceramic matrix, etc., and polymer reinforcements including laminar, particulate, and hybrid composites. This chapter also explores different matrices and reinforcements and their specific applications.

## 1.1 INTRODUCTION

Polymer composites are a novel family of materials that combine the unique characteristics of polymers with reinforcing elements. The idea for a composite material is for attaining certain physical and chemical attributes that other simple mono-component synthetic or natural materials cannot attain.[1] It can be used to create renewable resources, which are better than frequently used petroleum-based engineering polymers. The performance

Polymers and Functional Materials for a Cleaner Environment and Human Health: Sustainable Technologies and New Research. Ruby Varghese, Ann Rose Abraham, Shrikaant Kulkarni, & A. K. Haghi (Eds.)

gap between bio-based renewable polymers and petroleum polymers can be balanced out by combining them with bio-based renewable reinforcements.[2]

Polymer composite or PMC is a material that contains two components: matrix which holds the composite and reinforcement which acts as filler in the composite.[1] These composites can be either synthetic or natural. Polymers or polymers combined with various sorts of materials are used to create polymer composites. The performance of the resulting composite is improved by adding organic or inorganic fillers of a certain size and shape as reinforcement.[3,4]

The resulting substance has two parts as discussed earlier matrix and reinforcement. In a polymer composite, the polymer matrix is a continuous phase held in place by the reinforcing agents. They are joined macroscopically and are insoluble in one another.[1] Wood and bones are some examples of natural polymer composites.[5,6]

**Matrix:** The polymer composite is effectively held in place by the matrix. It could be elastomer, thermoplastic, thermosetting, or a combination of these. The matrix distributes the restrictions, connects reinforcing agents, ensures the structure's chemical resistance, and gives the resultant composite the correct shape. The choice of matrix used depends on the intended use of the polymer composite.[7]

**Reinforcement:** The reinforcement or reinforcing agent imparts strength and rigidity to the composite. It is the skeleton that provides mechanical strength to the polymer composites. The reinforcing agent typically consists of particles, strong fibers, or other materials that are embedded in the polymer matrix. While the polymer matrix holds everything in place reinforcement gives structure to the polymer.[7]

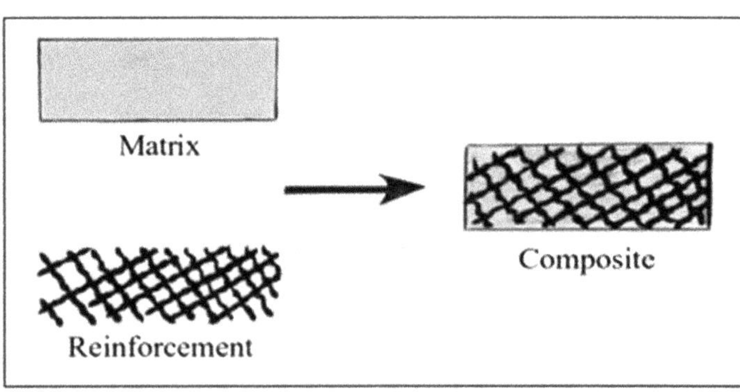


**FIGURE 1.1** Analogy of composite materials.

*Source*: Reprinted from Ref. [68]. https://creativecommons.org/licenses/by/4.0/

**Polymer Blend vs. Polymer Composite**

Polymer blend and polymer composite are both a mix of two or more polymers. The key difference between the two is that polymer blend is a single-phase system, whereas composite is a multiphase, multicomponent system, wherein each component's unique identities and characteristics are mirrored, resulting in a mix of mechanical properties that cannot be achieved from either component alone. Polymer composites have both chemical and physical bonds between polymer and non-polymer whereas polymer blends only form physical bonds.[1]

## 1.2 PROPERTIES OF POLYMER COMPOSITE

The properties of the polymer composite are determined by the characteristics of the polymer matrix and of the polymer reinforcement ratio in which they are combined and by the geometry and orientation of reinforcement in the composite.[1]

When the materials change to be a composite, their properties of strength, rigidity, and lightness change. Manufacturers can create traits precisely by choosing the proper combination of reinforcing and matrix materials for a particular purpose. They have several benefits, including biodegradability, low toxicity, and ease of supply, such as strong insulating properties, chemical qualities, and resilience to chemicals and the environment.[1]

## 1.3 CLASSIFICATION OF POLYMER COMPOSITES

They can be classified mainly into two groups: one based on the components of the polymer matrix, and the other based on the reinforcement form.

### *1.3.1 POLYMER MATRIX CONSTITUENTS*

#### *1.3.1.1 METAL MATRIX COMPOSITES (MMCS)*

A polymer composite in which a metal is used as a matrix is said to be a MMC. They have unique properties that make them preferable over monolithic metals for certain applications requiring higher specific strength, advanced specific modulus, enhanced performance at elevated temperatures, and lower coefficient of thermal expansion.[8] Based on the matrix material, they are further divided into composites based on Aluminum, Titanium, Magnesium, and Copper.

**Layered Metal Matrix Composites**

Layered MMCs are a unique category of laminate composites, created by stacking and bonding alternating layers of metal/alloy and MMCs. This arrangement significantly enhances the toughness and damage tolerance of the MMCs. Simulation and experimental tests have demonstrated that these layered composites exhibit greater ductility, toughness, and damage tolerance. This improvement can be attributed to essential phenomena such as crack bridging, deflections, blunting, and delamination.

Homogeneous reinforcement dispersion and excellent matrix adherence are the main factors that guarantee good composite performance. Ex-situ and in-situ fabrication processes can be separated based on the reinforcement and matrix reactions.[9]

#### *1.3.1.2 ALUMINUM-BASED COMPOSITES*

Aluminum metal matrix composites (AMCs) are broadly recognized for their compelling properties, especially for engineering applications.[10] Aluminum is either used as a cast alloy or a wrought alloy (AlZnMgCu, AlCuSiMn, AlMg, AlCu, AlMgSi, AlSiCuMg). Composites of Aluminum matrices have high specific strength and stiffness. They offer a combination of high strength and stiffness while being relatively lightweight. They have good wear and corrosion resistance, making them suitable for applications in challenging environments. They have a lower thermal expansion coefficient, making them less susceptible to dimensional changes with temperature variations and they retain their mechanical strength and properties even at elevated temperatures, making them suitable for applications requiring high-temperature performance.[11–13]

They are also notably cheap and easy to produce compared to other common MMCs.[14,15] AMC reinforced using industrial waste like fly ash, red mud and agricultural scraps like rice, husk, etc. are of current immense popularity.[16,17] AMCs have a perfect blend of material characteristics. They are inexpensive, easy to manufacture, and can incorporate waste materials as reinforcement. This makes them immensely popular as engineering materials.

### 1.3.1.3 TITANIUM-BASED COMPOSITES

Titanium alloys are excellent components for MMCs. They have high strength, stiffness, and a desirable strength-to-plasticity ratio. Traditional metallurgy is not applicable to titanium as they are highly chemically reactive. So titanium metal composites (TMCs) with discontinued filler-like particles or short fibers are widely produced using a process known as powder metallurgy (PM).[18–20] This also makes the process cost-effective and avoids material loss during metallurgy. Titanium and titanium alloys, such as dual-phase Ti-6Al-4V,[21] are one of the most often utilized matrices of MMCs. Pure titanium undertakes an allotropic conversion from hcp.

In titanium, the alloying elements are dissolved which can raise α-β transition temperature to stabilize the *α* phase (simple metals (SM) and most interstitial elements), lower the α-β transition temperature to stabilize the beta phase (transition metals (TM) and noble metals) and act only as a solid solution strengthener and not affect the transition temperature.

Based on these factors, titanium alloys are classified as three: *α* alloys, α+β alloys and β alloys. Which phases dominates room temperature and which does not, is dictated by the kind and quantity of alloying elements.[6,22] 6 wt% Aluminum stabilizing the α phase and 4 wt% vanadium stabilizing the β phase constitute an alloy, named Ti-6Al-4V alloy which is an example of α+β alloy. The structure is identified by the lamellar structure of α phase separated by β-phase interlayers inside the β-grain. The Ti-6Al-4V refers to α+β phase in a stable state containing 6% of the beta phase.

The microstructure is primarily composed of the hcp phase at room temperature, with some β phase (bcc) still present.[22] The alloys that fall under the α+β system includes one or more stabilizing or soluble elements in addition to one or more stabilizing elements. Wu. T et al. described the result of pre-deformation to the hot plasticity of the Ti-6Al-4V alloy.[23]

By in-situ processing, the titanium matrix's reactivity with elements like boron, carbon, and nitrogen can be explained. These composites perform better tribological performance as a result of the improved interfacial bonding shaped by in-situ techniques. In tribological studies, the wear, lubrication, and friction of contacting surfaces with respect to relative motion are captivated. The exceptional biocompatibility of titanium is a key characteristic and the alloys with low Young's modulus. β alloys are advantageous biomaterials for use in bone implants. The heat deformation process of the metastable Ti-6Cr-5Mo-5V-4Al alloy (Ti-6554) was analyzed by Liu et al.[24] It is regarded as a novel titanium alloy for producing aircraft structural parts as they have high strength and toughness. Zherebtsov et al. analysis of the spark plasma sintering produced the deformation behavior of Ti-15Mo/TiB composites.[25] They discovered that its increasing softening with an increase in deformation temperature between 500°C and 1000°C was connected to the dynamic retrieval and recrystallization methods. The reorientation and shortening of TiB whiskers were also reported. An extruded Ti-6.5Al-2Zr-1Mo-1V alloy (TA15) showed the tension-compression behavior at a temperature range of 650°C–750°C. This was detected by Zhang et al.[26]

**Powder Metallurgy**

Powder metallurgy (PM) is a manufacturing procedure that encompasses shaping and forming metal powders into various components and parts. This method leads to obtaining a uniformly distributed polymer matrix.

#### *1.3.1.4 MAGNESIUM-BASED COMPOSITES*

Magnesium composites are relatively a recent class of MMCs that are being utilized extensively in the automobile and aviation industries as they are light and have excellent vibration and damping properties. These composites have reduced density, strong mechanical characteristics, resilience against corrosion and wear, and a lower thermal expansion coefficient.

Magnesium is not adequately strong in its pure form, but it can be made to acquire specific properties like high specific strength when combined with other elements. Magnesium is used as a matrix in several nonferrous metals. Additionally, they are employed in applications where weight reduction is crucial.[8] Jayalakshmi et al. utilized squeeze-casting techniques to create magnesium composites.[27] Their findings revealed that the composite with the highest volume fraction exhibited a hardness value almost twice as much as the unreinforced base alloy, which had a hardness of 85 BHN. However, the unreinforced base alloy demonstrated a considerable decline in ultimate tensile strength as the test temperature increased, indicating its high susceptibility to temperature changes.

#### *1.3.1.5 COPPER-BASED METAL MATRIX COMPOSITES*

Copper-based MMCs provide great potential for use in electrical and electrotechnical applications such as heat exchangers, resistance welding electrodes, electronic connections, packaging, and more. To produce a superior mechanical property than pure copper, ceramic particles are added to the copper matrix. The most frequent strengthening phases are oxides ($Al_2O_3$, $Y_2O_3$, $SiO_2$), borides ($ZrB_2$, $TiB_2$), carbides (SiC, WC, TiC, and ZrC) and the use of others such as volcanic tuff, carbon, or intermetallic phases Al-Fe. [28–35]

Copper-based materials exhibit excellent electrical characteristics, elevated thermal conductivity, effective resilience against oxidation.[34–36] In copper matrices, homogeneously dispersed carbides are frequently used as reinforcement. These carbides play a crucial role in delaying the coarsening of particles during annealing processes, thereby the strength of the composite material was preserved. Tungsten carbide is a refractory material with a high melting point (2777°C).[36] High hardness (1000 kg/mm$^2$) at elevated temperatures outstanding oxidation resistance, and a high thermal stability.[37] The electrical contact in copper-tungsten carbide composite was primarily created using PM techniques such as powder blending, compacting, and sintering. PM may disseminate extremely tiny, reinforced particles into copper, improving microstructure control. With the improvement of composite density and hardness, Stobrawa and Rdzawski conducted a study showing that they successfully implanted fine and well-separated tungsten carbide particles into a copper matrix.[38] On the other hand, a lot of researchers are interested in PM in combination with mechanical alloying.[35,39,40]

High-energy milling is known to promote a phenomenon called mechanical alloying, which involves intense mechanical plastic deformation of the powders trapped in the milling jar. This deformation occurs due to collisions between the milling balls and the walls of the container, leading to the generation of many dislocations in the material. As a result, mechanical alloying enhances the structural and compositional homogeneity of the alloy and can lead to improved material properties. Sintering is a method for producing density-controlled materials by enhancing these solid-state processes. At a particular sintering temperature, when the green compact is completely densified entirely in a solid state, solid-state sintering takes place. To improve composite sintering ability, a variety of elements can be considered such as the green density, constituent element, particle size, heating rate, sintering temperature, sintering time, and sintering environment.

The mechanical alloying mechanism is used to promote the diffusion process of metals. To enhance the diffusion, sintering is helped by forming a new plane and so nanostructured alloys are formed. For composites made of copper, solid-state sintering typically takes place at temperatures between 800 and 1000°C.[41–43]

### *1.3.1.6 CERAMIC MATRIX COMPOSITES (CMCS)*

Ceramic matrix polymers have a matrix made of ceramic embedded between the reinforcing agents. Typical reinforcing agents include silicon carbide, carbon, mullite ($Al_6Si_2O_{13}$), and aluminum oxide. The main matrix components used are silicon carbide, zirconium oxide, and aluminum oxide.[44–48] CMC terminology is denoted as the principal fiber type/matrix type, carbon fibers reinforced silicon carbide is represented as C/silicon carbide. Even though ceramic materials habitually exhibit a combination of useful physical and mechanical properties, their applications are inadequate because of their brittle behavior. This can be overcome to a limit by using fiber reinforcements. It helps in improving the strength and toughness of the ceramic material. CMCs have a plethora of potential applications, but the mechanical and thermal stability of CMCs can be impacted by matrix microcracking. So, before such composites can be properly considered as useful substitutes for traditional materials, they must be thoroughly evaluated.[49]

CMCs are produced through industrial methods such as liquid phase sintering, hot pressing, and hot isostatic pressing. The presence of dispersed phase particles within the cracks formed in the ceramics enhances the material's toughness by interacting and arresting crack propagation. Numerous uses for CMCs are in the energy, aviation, and cutting tool industries. Examples of CMCs include borosilicate glass, silicon carbide (SiC), aluminum oxide ($Al_2O_3$), etc.

**Transformation Toughening**

Transformation toughening is a technique in which partially stabilized zirconia ($ZrO_2$) is dispersed into the matrix material to retain the metastable tetragonal phase at ambient conditions. The stress causes the particles to transform the monoclinic phase. If there is a slight increase in particle volume, producing compressive stresses near the tip's surface. This stops the growth of the particle.

### 1.3.1.7 BOROSILICATE GLASS

Borosilicate glass is a glass consisting of silica and boron trioxide as its major constituents. They typically have very low coefficients of thermal expansion ($5 \times 10^{-6}$ / °C at 20°C) and, therefore, they are more resistant to thermal shock than traditional glass.[50]

Glass/ceramic composites are reinforced with ceramic because of its superior mechanical and thermal properties, as well as its environmental friendliness and lower production costs.[51–54] On the other hand, glass has been extensively mixed with ceramics to produce composites that are appropriate for use as the substrate for microelectronics packaging. Because it has low thermal expansion and low dielectric constant for making ceramic circuits with high density, direct mounting of bare chips, and high-speed signal propagation at low firing temperatures. Additionally, using glass decreases the sintering temperature, making it possible to co-fire low-resistance conductors like gold and copper.[55]

### 1.3.1.8 SILICON CARBIDE

Silicon carbide (SiC) is made up of carbon and silicon in a tetrahedron structure held together by strong bonds in the crystal lattice. This produces a material that is incredibly tough and strong. Silicon carbide is resistant to acids, alkalis, and molten salts up to temperatures of 800°C and can withstand temperatures up to 1600°C in air. It forms a protective silicon oxide surface layer at 1200°C. Ceramics made of silicon carbide with little to no defects in their crystal structure can retain strength at extreme temperatures and have remarkable thermal shock resistance as they are strong thermal conductors with minimal thermal expansion and high strength. This material has a wide range of applications due to its diverse properties like high chemical purity, ability to maintain chemical inertness and strength in high temperatures, a good conductor of electricity, etc. Their uses range from wafer tray supports and paddles in semiconductor furnaces to thermistors (resistors that change resistance according to temperature), varistors (resistors that change resistance according to voltage), and resistive heating components in electrical furnaces.

It is used in several high-performance applications, including abrasives, refractories, and ceramics areas. Another application is in the field of electrical conductors, electronic components, and also in heat-resistance appliances.[50]

#### *1.3.1.9 ALUMINUM OXIDE*

Aluminum oxide ($Al_2O_3$), often referred to as alumina is an oxide of aluminum that is amphoteric in nature. It is a rather good conductor of heat for a CMC while being an electrical insulator. Aluminum oxides find applications in refractories, ceramics, and polishing and abrasive industries. In addition, large extents are utilized to produce zeolites, coat titanium pigments, act as a fire retardant and smoke suppressant.[50] The mechanical properties of the alumina fused CMCs are affected by their dense nano-crystalline structure, chemical composition, and grain size.[51,52]

### *1.3.2 POLYMER REINFORCEMENT CONSTITUENTS*

PMCs can be reinforced by natural or artificial fibers. They provide structure and impart unique properties to the composite. Reinforced polymer composites are now a popular replacement for conventional building materials like steel and concrete. Some of the popular fibers used in these composites are glass, carbon fiber, Kevlar, etc. Using natural fibers like flax, hemp, jute, sisal, kenaf, coir, kapok, and banana for these composites is also a research area as they have many beneficial properties that help in producing lightweight and economic polymer composites while also being environmentally friendly.[41,42] The reinforced fiber composite's mechanical characteristics can alter depending on a few parameters such as the reinforcing material's direction might be orthogonal, unidirectional, or random, the category of fibers, including carbon fiber and long-short glass fiber, and the fiber length and relative volume form.

### *1.3.2.1 GLASS FIBER REINFORCED POLYMER (GFRP) COMPOSITE*

GFRP composite or Fiber Reinforced Polymer (FRP) composites are formed using glass fibers embedded in a plastic matrix, enhancing their mechanical properties. They are also known as glass fiber-reinforced plastic. The glass fiber's diameter spans from 3 mm to 20 mm. Glass is frequently used as a reinforcing agent because of the following reasons such as ease of producing fiber from molten glass, economically inexpensive, adaptability to plenty of manufacturing techniques, high tensile strength, and chemical inertness. So, this composite is particularly useful in corrosive atmospheres. GFRP composites are principally used in transportation industries, plastic pipes, strong ampules, automotive and marine bodies, and industrial floorboards. After drawing new glass fibers, they are coated with a protective size, which shields the glass from adverse environmental conditions and potential damage. Before composite fabrication, the size is removed, and a coupling agent takes its place to enhance the bonding between the polymer matrix and the glass fiber. The coupling agent reinforces the bond between the polymer matrix and the fiber, improving their adhesion and overall performance in the composite material. These composites are not usually used in high-performance applications such as aviation or civil engineering despite having a high strength due to their of lack rigidity and stiffness.

### *1.3.2.2 CARBON FIBER REINFORCED POLYMER (CFRP) COMPOSITES*

In CFRP composites, the carbon fibers give the polymer materials rigidity and strength. The advantages of using these composites are, compared with all other categories of reinforcing fibers, carbon fibers have the upmost specific strength and specific modulus and can be engineered, carbon fibers can maintain their strength and modulus at high temperatures and at higher temperatures, they are not damaged by acids, bases, and moisture, the synthesis of carbon fibers is moderately cost-effective and reasonable, carbon fibers serve as a representation for non-crystalline and graphite areas. Pitch, rayon, and polyacrylonitrile (PAN) belong to the matrix materials, and epoxy size is often applied to carbon fibers, increasing the adherence of the polymer matrix.

Carbon fibers can be synthesized from polyacrylonitrile (PAN) and rayon. X-ray scattering and electron microscopy methods have been primarily used to investigate the microstructure of carbon fibers. Carbon fiber's structure is unlike that of graphite in that there is no three-dimensional order. During oxidative stabilization and subsequent carbonization, the linear chain structure of PAN-based fibers changes to a planar shape. Basal planes (a plane parallel to the lateral or horizontal axis) oriented along the fiber axis are formed during the carbonization stage.[56]

The major uses of CFRP composites are in military and commercial helicopters (wing, body, etc.), structural components for filament-wound rocket motor aviation structural components and pressure vessels, sports and recreational equipment (fishing rods, golf clubs, etc.), the manufacturing of accessories in the automobile industry.

#### *1.3.2.3 COMPOSITES REINFORCED WITH ARAMID FIBERS*

Aramid fiber is the first organic reinforcement to be used in PMC. Aramid fibers are more mechanically effective and equally weighty when compared to steel and glass fibers. This class of materials is popularly called Poly(p-phenylene) Terephthalamide and the most popular aramid materials are Nomex and Kevlar. They are similar materials but differ in the amide linkages on the aromatic ring. Kevlar fibers possess a wide range of desirable technical characteristics, including a high modulus, high strength, low weight, chemical inertness, and thermal stability. Although these fibers have great strength and moduli, they are poor in compression. Strong acids and bases can degrade these fibers; however, they can be used in high-temperature applications over temperatures ranging from -200°C to 200°C. Kevlar fibers are so often employed in flexible body armor as a functional material.[57] Nomex on the other hand is soluble in organic solvents and possesses greater elongation but it has a lower modulus and tensile strength than Kevlar. It possesses excellent flame resistance, dimensional stability, and chemical inertness to a variety of chemicals, while also having superior textile qualities making it a desirable product. The flexible polymer chain in Nomex provides it the textile qualities while maintaining high-temperature characteristics as of Kevlar.[58] In manufacturing items like car brakes, missile casings, bulletproof vests, confinement chambers for jet engines, cut-resistant gloves, sports goods, ropes, clutch lining gaskets, etc. are made up of aramid fiber-reinforced polymer composites.[59,60]

### *1.3.2.4 CARBON–CARBON COMPOSITES*

Polymer composites that use carbon fiber as both reinforcement and matrix are known as carbon-carbon composites. These composites have excellent tensile strength and moduli and are extremely resistant to thermal shocks. The presence of strong covalent link among the carbon-carbon atoms in them gives them extraordinary mechanical and thermal competences.[57,61–63] The thermal transport, strength, and modulus of these composites can be enhanced by utilizing fibers in the through-thickness direction and these composites show mechanical qualities such as extraordinary toughness and high strength at high temperatures exceeding 3000°C.[64] Carbon-carbon composites are mainly used in the manufacture of hot-pressed die, turbo jet engine components, X-ray targets, furnace fixturing, heat shields, load plates, heating elements, and various space structural applications. The components of C/C composites have been manufactured in the military, commercial aircraft, and defense applications with high-performance braking systems. The environmental applications such as atmospheric re-entry, solid rocket motor exhaust, and disk brakes in high-performance military and commercial air-crafts, high-speed trains, and racing cars.[45]

One of their primary drawbacks is that at higher temperatures, they are vulnerable to oxidation. Such composites require sophisticated processing methods, which drives up the cost of manufacture. This is so that carbon fibers may be impregnated with polymer resin, which is then allowed to cure to give it its final shape. They are preferred materials for demanding applications due to their exceptional mechanical properties relative to their weight and excellent resistance to high temperatures, making them ideal for severe and challenging conditions.

### *1.3.2.5 LAMINAR REINFORCEMENT*

Laminar composites are created through the joining of flat, two-dimensional sheets or panels, which are alternatively referred to as plies or laminae. Like continuous and aligned fiber-reinforced polymers, each ply in the laminate possesses a preferred high strength direction, resulting in a multi-layered structure. Two-dimensional laminae with various chosen high strength orientations are stalked and bonded together to form a laminar composite. Due to the enormous surface cross-sectional area to thickness ratios of the individual sheets, the phrase "two-dimensional" is employed. These sheets are frequently continuous, aligned composites with fiber reinforcement.

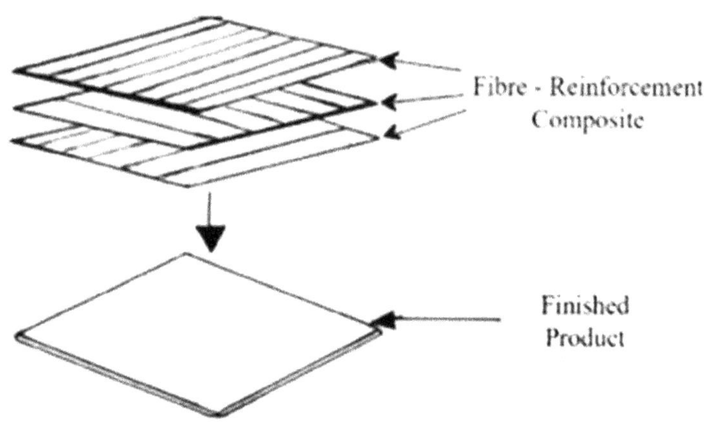


**FIGURE 1.2** Laminar reinforcement.

*Source*: Adapted from Ref. [67]

Layered composite materials offer an extensive range of qualities, each of which may be combined in a peculiar way, including high strength, corrosion resistance, and wear resistance.[47,48] Laminar composites can be classified as unidirectional composites, where all laminae are in a unidirectional structure and have the same orientation for the high strength direction as shown in Figure 1.3(a). Cross-ply composites, where the presence of alternate 0° and 90° orientations give high strength layers as shown in Figure 1.3(b). Angle-ply composites, where the consecutive layers alternate between favorable and unfavorable orientations characterized by high strength (e.g. 145°) as shown in Figure 1.3(c). Multidirectional composites, where the laminae are arranged in diverse high strength orientations as shown in Figure 1.3(d). In Figure 1.3(a) to (d) laminar composite lay-ups (schematics) are depicted.[49,65]

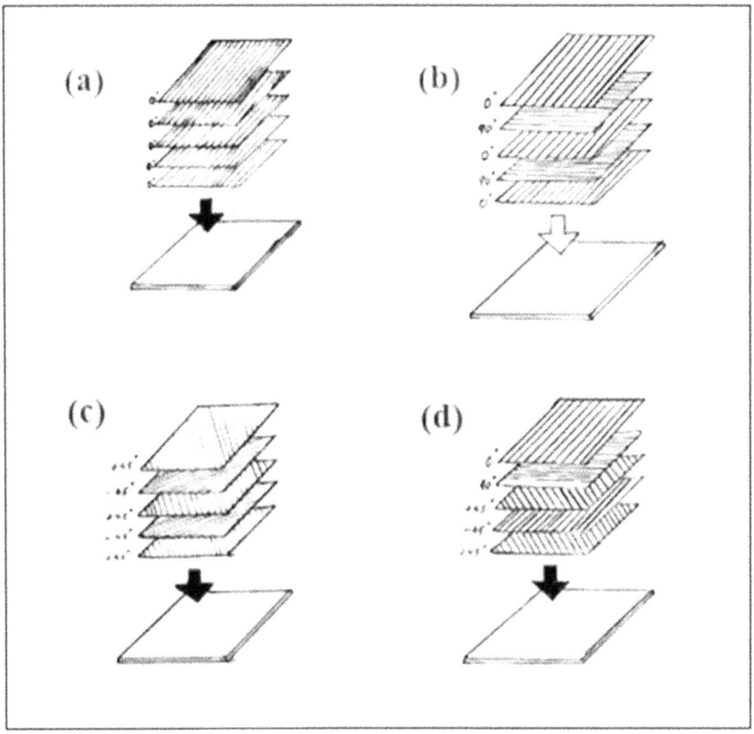


**FIGURE 1.3** (a) Unidirectional laminate, (b) cross-ply laminate, (c) angle-ply laminate, and (d) multidirectional laminate.

In most laminates, the layers are arranged symmetrically with respect to the laminate midplane, ensuring that fiber orientations are balanced and avoiding any out-of-plane twisting or bending. Unidirectional laminates have extremely anisotropic in-plane qualities, such as elasticity and strength.[43] The degree of in-plane isotropy is increased by cross, angle, and angle-ply materials.

Sandwich panels, a subclass of structural composites, were created to be lightweight beams or panels with excellent stiffness and strength levels. It is made up of a thicker core that is adherently joined to two exterior sheets, faces, or skins that are spaced apart from one another. Because the materials used for the sandwich panel's core are less expensive than those used for its faces, it is a cost-effective composite. The main components used in them are hard polymeric foams, wood, and honeycombs. They are used in the industries of aviation, automobiles, ships, and construction and civil infrastructure and in the manufacture of automotive components including auto panels, sports car bodywork, and drive shafts; naval components like ship hulls, hatch covers, deckhouses, bulkheads, and propellers. They are also desirable in building civil infrastructures like bridge components, long-span roof structures, beams, structural panels, roof panels, and tanks.

Honeycomb structures are composed of slender foils formed into interconnected cells, frequently organized in hexagonal or alternative patterns, with axes oriented perpendicular to the surface planes. Arrangements in this manner, featuring axes that are perpendicular to the surface planes, lead to structures displaying mechanical characteristics that are anisotropic. These structures exhibit their highest resistance to shear forces within the plane of the panel. However, their ability to withstand tensile and compressive forces is most prominent when those forces align parallel to the axis of the individual cells that make up the panel.

The strength and stiffness of honeycomb structures are influenced by the cell size, thickness of the cell walls, and the material used. Furthermore, the significant proportion of vacant space within each cell contributes to exceptional sound absorption and vibration-damping capabilities. Honeycomb structures are formed by employing thin sheets of materials such as metal alloys and polymers to shape the core frameworks.

#### *1.3.2.6 PARTICULATE REINFORCEMENT*

Another kind of reinforcement used in composite materials, particulate reinforcement has an isotropic structure and composition.[51–53] Here the primary constituents, are homogeneously spread all over the bulk material.[56]

In these types of composites, particles are suspended in the polymer matrix. Depending on the intended application these particles can have any range of size, shape, or arrangement. Concrete and particle boards are two common types of particulate composites. These composites can be subclassified based on particulates used, into flakes and filled/skeletal. A flake composite commonly consists of suspended flakes within a matrix material, like particle board, characterized by a notable ratio of platform area to thickness. Filled/Skeletal: this is made up of a continuous skeletal matrix filled with a second material, such as an insulating substance in the case of a honeycomb core. An anisotropic or orthotropic response can be seen in a particle composite. Such composites are employed in several applications where strength is not a key consideration in the design.[54]

Materials exhibiting distinct and independent characteristics in the three orthogonal axes are said to be orthotropic, examples are crystals, wood, rolled metals, etc.

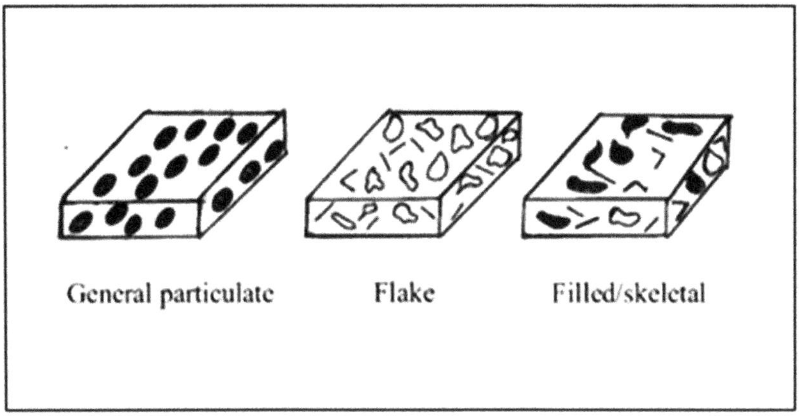


**FIGURE 1.4** A schematic of different types of particulate composites.

*Source*: Adapted from Ref. [55]

Particulate reinforced composites are applicable in situations where high levels of wear resistance are required, such as on road surfaces. Gravel is used in paving roads as a reinforcing filler, to increase the hardness of cement significantly. Moreover, particulate composites find application in producing magnets, dental fillings, brakes, darts, bio-implants, and cutting tools.

**Hybrid Composites**

Hybrid composite materials are made by fusing at least two reinforcement elements into a matrix to improve the properties of the composite. The various pieces' shape, composition, and orientation have a significant impact on how composites respond to stimuli. The attributes of the composite are closely tied to the traits of its constituent parts, which control its internal structures.[66]

## KEYWORDS


- **polymer matrix composites**
- **aluminum-based composites**
- **titanium-based composites**
- **ceramic matrix**
- **glass fiber-reinforced polymer composites**
- **laminar reinforcement**


## REFERENCES


1. Kulshreshtha, A. K.; Vasile, C. *Handbook of Polymer Blends and Composites*, Vol. 1; iSmithers Rapra Publishing, 2002.
2. Ruzuqi, R. Impact Strength Analysis of Polymer Composite Materials (PCM) Fiber Reinforced in the Fiberboat Application. *Mater. Sci. Res. India* **2020,** *17* (2), 170–178. https://doi.org/10.13005/msri/170209.
3. Thomas, S.; Joseph, K.; Malhotra, S. K.; Goda, K.; Sreekala, M. S. *Polymer Composites, Macro-and Microcomposites*, Vol. 1; John Wiley & Sons, 2012.
4. Meijer, H. E.; Govaert, L. E. Mechanical Performance of Polymer Systems: The Relation Between Structure and Properties. *Prog. Polym. Sci.* **2005,** *30* (8–9), 915–938.
5. Freudenberg, K. The Relation of Cellulose to Lignin in Wood. *J. Chem. Educ.* **1932,** *9* (7), 1171.
6. Lee, K.-Y.; Buldum, G.; Mantalaris, A.; Bismarck, A. More Than Meets the Eye in Bacterial Cellulose: Biosynthesis, Bioprocessing, and Applications in Advanced Fiber Composites. *Macromol. Biosci.* **2014,** *14* (1), 10–32.
7. Hsissou, R.; Seghiri, R.; Benzekri, Z.; Hilali, M.; Rafik, M.; Elharfi, A. Polymer Composite Materials: A Comprehensive Review. *Compos. Struct.* **2021,** *262*, 113640.
8. Dash, D.; Samanta, S.; Rai, R. Study on Fabrication of Magnesium Based Metal Matrix Composites and Its Improvement in Mechanical and Tribological Properties-a Review. In: *IOP Conference Series: Materials Science and Engineering*, Vol. 377; IOP Publishing, 2018; p 012133.
9. Kondoh, K. Titanium Metal Matrix Composites by Powder Metallurgy (PM) Routes. In: *Titanium Powder Metallurgy*; Elsevier, 2015; pp 277–297.
10. Senthilvelan, T.; Gopalakannan, S.; Vishnuvarthan, S.; Keerthivaran, K. Fabrication and Characterization of SiC, $Al_2O_3$ and $B_4C$ Reinforced Al-Zn-Mg-Cu Alloy (AA 7075) Metal Matrix Composites: A Study. *Adv. Mater. Res.* **2013,** *622*, 1295–1299.
11. Hao, S.; Xie, J. Tensile Properties and Strengthening Mechanisms of SiCp-Reinforced Aluminum Matrix Composites as a Function of Relative Particle Size Ratio. *J. Mater. Res.* **2013,** *28* (15), 2047–2055.

12. Sajjadi, S. A.; Ezatpour, H. R.; Beygi, H. Microstructure and Mechanical Properties of Al–Al2O3 Micro and Nano Composites Fabricated by Stir Casting. *Mater. Sci. Eng. A* **2011,** *528* (29–30), 8765–8771.
13. Kaveendran, B.; Wang, G.; Huang, L.; Geng, L.; Peng, H. In Situ (Al3Zr+ Al2O3np)/2024Al Metal Matrix Composite with Novel Reinforcement Distributions Fabricated by Reaction Hot Pressing. *J. Alloys Compd.* **2013,** *581*, 16–22.
14. Rohatgi, P.; Schultz, B. Lightweight Metal Matrix Nanocomposites–Stretching the Boundaries of Metals. *Mater. Matters* **2007,** *2* (4), 16–20.
15. Tahamtan, S.; Halvaee, A.; Emamy, M.; Zabihi, M. Fabrication of Al/A206–Al2O3 Nano/Micro Composite by Combining Ball Milling and Stir Casting Technology. *Mater. Des.* **2013,** *49*, 347–359.
16. Prasad, N.; Acharya, S. Development and Characterization of Metal Matrix Composite Using Red Mud an Industrial Waste for Wear Resistant Applications. *Development* **2006**.
17. Alaneme, K. K.; Adewale, T. M.; Olubambi, P. A. Corrosion and Wear Behaviour of Al–Mg–Si Alloy Matrix Hybrid Composites Reinforced with Rice Husk Ash and Silicon Carbide. *J. Mater. Res. Technol.* **2014,** *3* (1), 9–16.
18. Kumar, P.; Chandran, K. R. Strength–Ductility Property Maps of Powder Metallurgy (PM) Ti-6Al-4V Alloy: A Critical Review of Processing-Structure-Property Relationships. *Metall. Mater. Trans. A* **2017,** *48*, 2301–2319.
19. Konstantinov, A.; Bazhin, P.; Stolin, A.; Kostitsyna, E.; Ignatov, A. Ti-B-Based Composite Materials: Properties, Basic Fabrication Methods, and Fields of Application. *Compos. Part Appl. Sci. Manuf.* **2018,** *108*, 79–88.
20. Hayat, M. D.; Singh, H.; He, Z.; Cao, P. Titanium Metal Matrix Composites: An Overview. *Compos. Part Appl. Sci. Manuf.* **2019,** *121*, 418–438.
21. Rosso, M. Ceramic and Metal Matrix Composites: Routes and Properties. *J. Mater. Process. Technol.* **2006,** *175* (1–3), 364–375.
22. Semiatin, S.; Seetharaman, V.; Weiss, I. Hot Working of Titanium Alloys–an Overview. *Adv. Sci. Technol. Titan. Alloy Process.* **1996,** 3–73.
23. Wu, T.; Wang, N.; Chen, M.; Zuo, D.; Xie, L.; Shi, W. Effect of Pre-Strain on Microstructure and Tensile Properties of Ti-6al-4v at Elevated Temperature. *Metals* **2021,** *11* (8), 1321.
24. Liu, Q.; Wang, Z.; Yang, H.; Ning, Y. Hot Deformation Behavior and Processing Maps of Ti-6554 Alloy for Aviation Key Structural Parts. *Metals* **2020,** *10* (6), 828.
25. Zherebtsov, S.; Ozerov, M.; Klimova, M.; Moskovskikh, D.; Stepanov, N.; Salishchev, G. Mechanical Behavior and Microstructure Evolution of a Ti-15Mo/TiB Titanium–Matrix Composite during Hot Deformation. *Metals* **2019,** *9* (11), 1175.
26. Zhang, C.; Li, D.; Li, X.; Li, Y. An Experimental Study of the Tension-Compression Asymmetry of Extruded Ti-6.5 Al-2Zr-1Mo-1V under Quasi-Static Conditions at High Temperature. *Metals* **2021,** *11* (8), 1299.
27. Jayalakshmi, S.; Kailas, S. V.; Seshan, S. Tensile Behaviour of Squeeze Cast AM100 Magnesium Alloy and Its $Al_2O_3$ Fibre Reinforced Composites. *Compos. Part Appl. Sci. Manuf.* **2002,** *33* (8), 1135–1140.
28. Moustafa, S.; Abdel-Hamid, Z.; Abd-Elhay, A. Copper Matrix SiC and $Al_2O_3$ Particulate Composites by Powder Metallurgy Technique. *Mater. Lett.* **2002,** *53* (4–5), 244–249.

29. Celebi Efe, G.; Altinsoy, I.; İpek, M.; Zeytin, S.; Bindal, C. Effects of SiC Particle Size on Properties of Cu-SiC Metal Matrix Composites. *Acta Phys. Pol. A* **2012,** *121* (1), 251–253.
30. Yusoff, M.; Hussain, Z. Effect of Sintering Parameters on Microstructure and Properties of Mechanically Alloyed Copper-Tungsten Carbide Composite. *Int. J. Mater. Mech. Manuf.* **2013,** *1* (3), 283–286.
31. Vu, H.; Tran, L.; Duong, B.; Dang, K. Consolidation and Properties of Cu-TiC Composite by Reduction Sintering and Cold Extrusion Process. In: *ECCM15-15TH European Conference on Composite Materials*, Venice, Italy; 2012.
32. Chi, F.; Schmerling, M.; Eliezer, Z.; Marcus, H.; Fine, M. Preparation of Cu-TiN Alloy by External Nitridation in Combination with Mechanical Alloying. *Mater. Sci. Eng. A* **1995,** *190* (1–2), 181–186.
33. Mikula, J.; Lach, M. Kompozyt Miedź-Tuf Wulkaniczny: Wytwarzanie, Wlasności i Zastosowania. *Czas. Tech. Mech.* **2011,** *108* (3–M), 53–60.
34. Lee, D.; Kim, B. Nanostructured Cu–Al2O3 Composite Produced by Thermochemical Process for Electrode Application. *Mater. Lett.* **2004,** *58* (3–4), 378–383.
35. Lopez, M.; Corredor, D.; Camurri, C.; Vergara, V.; Jimenez, J. Performance and Characterization of Dispersion Strengthened Cu–$TiB_2$ Composite for Electrical Use. *Mater. Charact.* **2005,** *55* (4–5), 252–262.
36. Stobrawa, J.; Rdzawski, Z. Characterisation of Nanostructured Copper-WC Materials. *J. Achiev. Mater. Manuf. Eng.* **2009,** *32* (2), 171–178.
37. Cardarelli, F. *Materials Handbook: A Concise Desktop Reference*; 2008.
38. J, S.; Rdzawski, Z.; Głuchowski, W.; Malec, W. Microstructure and Properties of CuNi2Si1 Alloy Processed by Continuous RCS Method. *J. Achiev. Mater. Manuf. Eng.* **2009,** *37*.
39. Besterci, M.; Kovac, L. Microstructure and Properties of Cu-$Al_2O_3$ Composites Prepared by Powder Metallurgy. *Int. J. Mater. Prod. Technol.* **2003,** *18* (1–3), 26–56.
40. Salahinejad, E.; Amini, R.; Marasi, M.; Hadianfard, M. J. The Effect of Sintering Time on the Densification and Mechanical Properties of a Mechanically Alloyed Cr–Mn–N Stainless Steel. *Mater. Des.* **2010,** *31* (1), 527–532.
41. Stobrawa, J.; Rdzawski, Z.; Gluchowski, W. Microstructure and Properties of Nanocrystalline Copper-Yttria Microcomposites. *J. Achiev. Mater. Manuf. Eng.* **2007,** *24* (2), 83–86.
42. Stobrawa, J.; Rdzawski, Z. Dispersion–Strengthened Nanocrystalline Copper. *J. Achiev. Mater. Manuf. Eng.* **2007,** *24* (2), 35–42.
43. Gan, K.; Gu, M. Mechanical Alloying Process and Mechanical Properties of Cu–SiCp Composite. *Mater. Sci. Technol.* **2006,** *22* (8), 960–964.
44. Baxevanis, T.; Charalambakis, N. A Micromechanically Based Model for Damage-Enhanced Creep-Rupture in Continuous Fiber-Reinforced Ceramic Matrix Composites. *Mech. Mater.* **2010,** *42* (5), 570–580.
45. Choi, S. R.; Kowalik, R. W.; Alexander, D. J.; Bansal, N. P. Elevated-Temperature Stress Rupture in Interlaminar Shear of a Hi-Nic SiC/SiC Ceramic Matrix Composite. *Compos. Sci. Technol.* **2009,** *69* (7–8), 890–897.
46. Sommers, A.; Wang, Q.; Han, X.; T'Joen, C.; Park, Y.; Jacobi, A. Ceramics and Ceramic Matrix Composites for Heat Exchangers in Advanced Thermal Systems—A Review. *Appl. Therm. Eng.* **2010,** *30* (11–12), 1277–1291.

47. Fantozzi, G.; Reynaud, P. Mechanical Hysteresis in Ceramic Matrix Composites. *Mater. Sci. Eng. A* **2009,** *521*, 18–23.
48. Pompidou, S.; Lamon, J. Analysis of Crack Deviation in Ceramic Matrix Composites and Multilayers Based on the Cook and Gordon Mechanism. *Compos. Sci. Technol.* **2007,** *67* (10), 2052–2060.
49. Donald, I. W.; McMillan, P. W. Ceramic-Matrix Composites. *J. Mater. Sci.* **1976,** *11*, 949–972.
50. Park, S.-J.; Seo, M.-K. Comprehension of Nanocomposites. In: *Interface Science and Technology*, Vol. 18; Elsevier, 2011; pp 777–819.
51. Schmücker, M.; Mechnich, P. Microstructural Coarsening of Nextel™ 610 Fibers Embedded in Alumina-Based Matrices. *J. Am. Ceram. Soc.* **2008,** *91* (4), 1306–1308.
52. Wilson, D. M.; Lieder, S.; Lueneburg, D. Microstructure and High Temperature Properties of Nextel 720 Fibers. In: *Proceedings of the 19th Annual Conference on Composites, Advanced Ceramics, Materials, and Structures—B: Ceramic Engineering and Science Proceedings*; Wiley Online Library, 1995; pp 1005–1014.
53. Saheb, D. N.; Jog, J. P. Natural Fiber Polymer Composites: A Review. *Adv. Polym. Technol. J. Polym. Process. Inst.* **1999,** *18* (4), 351–363.
54. Manna, S.; Saha, P.; Chowdhury, S.; Thomas, S.; Sharma, V. Alkali Treatment to Improve Physical, Mechanical and Chemical Properties of Lignocellulosic Natural Fibers for Use in Various Applications. *Lignocellul. Biomass Prod. Ind. Appl.* **2017,** 47–63.
55. Staab, G. H. 1—Introduction to Composite Materials. In: *Laminar Composites*; Staab, G. H., Ed.; Butterworth-Heinemann: Woburn, 1999; pp 1–16. https://doi.org/10.1016/B978-075067124-8/50001-1.
56. Henkel, D.; Pense, A. W. Structure and Properties of Engineering Materials. *No Title* **2002**.
57. Florea, R. M.; Carcea, I. Polymer Matrix Composites–Routes and Properties. *Int. J. Mod. Manuf. Technol.* **2012,** *4* (1), 59–64.
58. Kutz, M. *Handbook of Materials Selection*; John Wiley & Sons, 2002.
59. Sapuan, S. M. A Computer-Aided Material Selection for Design of Automotive Safety Critical Components with Novel Materials. *Malays. J. Comput. Sci.* **1999,** *12* (2), 37–46.
60. Yaacob, A. M.; Ahmad, M.; Dahlan, K. Z. M.; Sapuan, S. M. Experimental Studies on Fiber Orientation of Short Glass Fiber Reinforced Injection Molded Thermoplastic Composites. In: *Proceedings of Advanced Materials Conference, Advanced Technology Congress, Putra Jaya*; 2003; pp 20–21.
61. Asthana, R.; Kumar, A.; Dahotre, N. B. *Materials Processing and Manufacturing Science*; Elsevier, 2006.
62. Nedelcu, D.; Carcea, I. Technology for Obtaining Samples of Layered Composite Materials with Metallic Matrix. *Met. Mater. Int.* **2013,** *19*, 105–112.
63. Lee, K.-Y.; Aitomäki, Y.; Berglund, L. A.; Oksman, K.; Bismarck, A. On the Use of Nanocellulose as Reinforcement in Polymer Matrix Composites. *Compos. Sci. Technol.* **2014,** *105*, 15–27.
64. Tang, Q.; Wu, J.; Sun, H.; Fan, S.; Hu, D.; Lin, J. Superabsorbent Conducting Hydrogel from Poly (Acrylamide-Aniline) with Thermo-Sensitivity and Release Properties. *Carbohydr. Polym.* **2008,** *73* (3), 473–481.

65. Park, S.-J.; Seo, M.-K. *Interface Science and Composites*, Vol. 18; Academic Press, 2011.
66. Bouhfid, N.; Raji, M.; Boujmal, R.; Essabir, H.; Bensalah, M.-O.; Bouhfid, R. et al. Numerical Modeling of Hybrid Composite Materials. In: *Modelling of Damage Processes in Biocomposites, Fibre-Reinforced Composites and Hybrid Composites*; Elsevier, 2019; pp 57–1
67. Egbo, M. A Fundamental Review on Composite Materials and Some of Their Applications in Biomedical Engineering. *J. King Saud Univ. Eng. Sci.* 2020, 33. 10.1016/j.jksues.2020.07.007.
68. Ruzuqi R. Impact Strength Analysis of Polymer Composite Materials (PCM) Fiber Reinforced in the Fiberboat Application. *Mat. Sci. Res. India* 17(2).

# CHAPTER 2

# Sustainable Polymers for Healthcare

NEENU GEORGE[1], UNNIKRISHNAN T. G.[2], JITHIN JOY[3], and CINTIL JOSE[3]

[1]*Department of Chemistry, St. Joseph's College, Moolamattom, Idukki, Kerala, India*

[2]*Department of Mechanical Engineering, Viswajyothi College of Engineering and Technology, Kerala, India*

[3]*Department of Chemistry, Newman College, Thodupuzha, Idukki, Kerala, India*

## ABSTRACT

This chapter delves into the diverse and evolving landscape of biomedical applications of polymers, encompassing both synthetic and natural base biopolymers. Polymers have emerged as versatile materials with wide-ranging utility in the biomedical field, addressing various therapeutic and diagnostic needs. In hard tissue applications, polymers play a pivotal role in orthopedics, dental implants, and bone tissue engineering, offering biocompatibility, mechanical strength, and controlled drug delivery. Soft tissue applications explore the potential of polymers in wound healing, tissue scaffolds, and drug release systems, with a focus on promoting tissue regeneration and minimizing inflammation. Beyond these traditional domains, polymers find innovative uses in shape-memory polymers for biomedical applications, allowing for adaptive devices and controlled drug release in response to specific triggers. Porous organic polymers (POPs) are revolutionizing drug delivery and bioseparation techniques,

Polymers and Functional Materials for a Cleaner Environment and Human Health: Sustainable Technologies and New Research. Ruby Varghese, Ann Rose Abraham, Shrikaant Kulkarni, & A. K. Haghi (Eds.)

providing controlled release kinetics and targeted therapy. Meanwhile, natural base biopolymers continue to gain prominence in drug delivery, wound dressings, and tissue engineering due to their biodegradability and minimal immunogenicity. This chapter discusses recent advancements, challenges, and future prospects, providing a comprehensive overview of the significant role polymers play in revolutionizing healthcare and improving patient outcomes.

## 2.1 INTRODUCTION

Polymers have emerged as versatile materials with profound implications in the field of biomedical applications. These synthetic or natural compounds have demonstrated their utility in various facets of healthcare, offering innovative solutions for diverse medical needs.[1,2] This introduction will provide an overview of the biomedical applications of polymers, highlighting their significance in hard tissue and soft tissue applications, exploring other medical uses, delving into the promising field of shape-memory polymers, and discussing the role of POPs and natural-based biopolymers in the realm of biomedicine.

One of the pivotal areas where polymers have made a significant impact is hard tissue applications. This includes the development of polymer-based biomaterials for bone repair, dental prostheses, and orthopedic implants.[3] The biocompatibility and tunable mechanical properties of polymers have allowed for the creation of implants that can integrate seamlessly with the surrounding hard tissues, promoting faster healing and improved patient outcomes.[4] In the realm of soft tissue applications, polymers have also proven invaluable. These materials have been employed in the development of tissue scaffolds, wound dressings, and drug delivery systems, among others. Their biocompatibility, flexibility, and ability to mimic the extracellular matrix make polymers suitable for promoting tissue regeneration and wound healing.[5] Beyond hard and soft tissue applications, polymers have found utility in various other medical fields. They are used in medical devices such as catheters, sutures, and contact lenses.[6] Polymer-based nanoparticles and microparticles are employed for targeted drug delivery and diagnostic purposes. Additionally, polymers play a crucial role in tissue engineering, organ transplantation, and even in the development of controlled release vaccines.

Shape-memory polymers represent an exciting frontier in biomedical research. These polymers have the unique ability to "remember" their original shape and return to it upon the application of specific stimuli.[7] In the medical field, shape-memory polymers are being explored for minimally invasive surgical tools, stents, and drug delivery devices, where their controllable shape-changing properties offer innovative solutions.[8] Natural-based biopolymers, derived from sources such as collagen, chitosan, and alginate, have garnered interest in the biomedical field. Their biodegradability and biocompatibility make them suitable for wound healing, tissue engineering, and drug delivery applications.[9]

In summary, this book chapter discusses the application of polymers in biomedicine that spans a wide range of areas, from hard tissue and soft tissue applications to diverse medical uses, including emerging technologies like shape-memory polymers and the innovative use of porous organic and natural-based biopolymers. This dynamic field continues to evolve, offering promising solutions to some of the most challenging issues in healthcare.

## 2.2 HARD TISSUES

Recent studies have highlighted the significance of polymer-based scaffolds for bone tissue engineering. Researchers have been exploring advanced polymers, including biodegradable ones like polylactic acid (PLA) and polyglycolic acid (PGA), to develop scaffolds that mimic the natural extracellular matrix (ECM).[10–12] These scaffolds promote cell attachment, proliferation, and differentiation, making them ideal for bone regeneration applications. Hydrogels, which are highly water-absorbent polymers, have gained attention for their potential in cartilage repair. Recent research has focused on developing injectable hydrogels that can be used for minimally invasive procedures. These hydrogels provide a supportive environment for chondrocyte growth and cartilage formation, offering promising solutions for osteoarthritis treatment.[13]

Latest research has explored the use of biopolymer coatings on orthopedic implants to improve osseointegration.[14] These coatings, often composed of biocompatible polymers like hyaluronic acid or chitosan, can enhance the integration of implants with surrounding bone tissue.[15] This research aims to reduce the risk of implant-related complications and promote faster healing. Advances in 3D printing technology have enabled the fabrication of patient-specific polymer scaffolds for bone implants. Recent studies have demonstrated the feasibility of creating intricate and customized implants using biocompatible polymers like polyethylene, enabling a tailored approach to bone defect reconstruction.[16]

In recent research, polymer-ceramic composites have been investigated for their potential in hard tissue applications.[17] These composites combine the strength and biocompatibility of polymers with the bioactivity of ceramics, such as hydroxyapatite. The result is a material with improved mechanical properties and osteoconductivity, making it suitable for load-bearing applications.[17,18]

## 2.3 SOFT TISSUES

Recent studies have emphasized the use of biodegradable polymers, such as poly(lactic-co-glycolic acid) (PLGA) and polyurethanes, for soft tissue engineering applications. These polymers offer tunable degradation rates and have been employed in the development of scaffolds for applications like skin regeneration and vascular tissue engineering.[19] Hydrogels based on polymers like poly(ethylene glycol) (PEG) and poly(acrylic acid) (PAA) have gained prominence in wound dressing applications. Recent research has focused on creating smart and responsive hydrogels that can promote wound healing by maintaining a moist environment and releasing therapeutic agents as needed.[20]

Polymers have been extensively used to fabricate nanoparticles for drug delivery to soft tissues. Recent work has centered on designing nanoparticles with controlled release properties, high drug loading capacity, and the ability to target specific soft tissue sites, such as tumors, to enhance therapeutic efficacy and minimize side effects.[21] Polymeric sutures, including absorbable materials like polyglycolide (PGA) and polylactide (PLA), have been used for soft tissue repair.[22] Recent research has focused on enhancing the biomechanical properties and biocompatibility of these sutures, making them suitable for applications in cardiovascular surgery and tissue reconstruction.[23] Polymer-based nerve conduits have shown promise in facilitating peripheral nerve repair. Recent research has explored the use of polymers like polycaprolactone (PCL) and poly(l-lactic acid) (PLLA) to create conduits that guide and support nerve regeneration, offering potential solutions for injuries and peripheral neuropathies.[22]

## 2.4 OTHER MEDICAL APPLICATIONS

Polymers are integral components of various medical devices, including catheters, stents, and prosthetic heart valves. Recent research has focused on enhancing the biocompatibility, durability, and functionality of these devices through the use of advanced polymer materials. Polymer nanoparticles have gained attention as carriers for targeted drug delivery.[24] Recent studies have explored the design and optimization of polymer-based nanoparticles, such as polymeric micelles and nanogels, for the controlled release of drugs in various medical conditions. Shape-memory polymers have been utilized in the development of minimally invasive medical devices.[25] Recent research has focused on designing polymers that can undergo controlled shape changes in response to external stimuli, enabling applications such as stents that can be delivered through small incisions and expand once in position. Polymers have found applications in organ preservation solutions used in transplantation procedures.[26] Recent studies have explored the development of polymer-based solutions that can better maintain the viability of donor organs during storage, potentially extending the time available for transplantation. Polymeric materials have been investigated for their potential in developing controlled release vaccines and drug delivery systems.[26] Recent research has focused on designing polymer-based carriers that can deliver vaccines or therapeutic agents with improved efficacy and stability.[27]

## 2.5 SHAPE-MEMORY POLYMERS

Shape-memory polymers (SMPs) are a class of polymers that exhibit the unique ability to transition between a permanent shape and a temporary shape, with the potential for repeated shape changes. The temporary shape is achieved mechanically deforming the permanent shape under particular conditions, such as heating above the glass transition or melting temperature, followed by fixing of the temporary shape through methods like chemical crosslinking, crystallization, or supramolecular interactions. Throughout the duration of the deformation, the polymer keeps its temporary shape until an appropriate trigger, such as light, an increase in temperature, or contact with particular chemicals, is applied to commence the recovery of the permanent shape. While temperature and light have been the primary direct triggers documented in the literature, indirect triggers, including exposure to solvents, have also been explored for their potential in shape recovery,[28,29] ultrasound,[3] and others have also been reported.[30,31]

In recent decades, there has been a growing focus on shape-memory materials within the field of biomedical research.[32–34] SMPs, with their unique capability to retain complex shapes distinct from their permanent form, offer promising potential in the development of tissue-engineered constructs.[29] In fact, they make minimally invasive surgery possible by featuring a compact temporary structure capable of expanding to its ultimate (patient-specific) shape during in vivo implantation.[35,36] Furthermore, they empower the automatic execution of intricate mechanical deformations that would otherwise require manual manipulation by a surgeon.[37] To demonstrate the potential of this category of materials, an early record highlighting SMPs was a 1941 patent describing an elastic polymer resin "elastic memory" intended for use in dental applications. This resin was composed of (m) ethyl methacrylate, vinyl chloride, vinyl acetate, and copolymers, as described in the patent.[38] Subsequently, the exploration of SMPs that may have biomedical benefits has garnered escalating attention.

### 2.5.1 BIOMEDICAL APPLICATIONS OF SHAPE-MEMORY POLYMERS

Polymers with shape-memory have already been developed and found in numerous applications, encompassing both biomedical and nonbiomedical domains (e.g., heat shrink tubes). However, this review will specifically focus on the biomedical applications of SMPs.

#### 2.5.1.1 SHAPE-MEMORY POLYMERS FOR DRUG DELIVERY

The integration of drug administration that remembers its shape presents an intriguing opportunity, as the shape-memory effect can effortlessly disable the drug delivery system at the intended drug release site without requiring external manipulation. This feature enables long-lasting drug delivery at the target site and eliminates the need for repeated injections. or surgeries. Notably, Wischke et al. have shown that drug loading in a copolyester urethane shape-memory polymer increases drug release and has minimal impact on the material's shape-memory capabilities. Their study further illustrates the effective application of shape-memory materials for the discharge of drugs in aqueous environments. Remarkably, the study highlights that the incorporation of drugs does not significantly alter the polymer's glass transition and elastic properties when exposed to the aqueous environment. The researchers anticipate the utilization of these polymers for the controlled release of low-drug doses over several months.[39,40] Nevertheless, a recent publication by Jahangiri et al. revealed that the loading of drugs through a swelling method had a detrimental effect on shape fixity values compared to an in situ drug loading approach. The study demonstrated a 23% decrease in shape fixity after the drug was released, highlighting the contrasting outcomes between the two loading methods.[41] These findings indicate that the performance of drug-loaded SMPs is highly influenced by the method of drug incorporation. To preserve the shape-memory performance, researchers should prioritize the use of an in situ drug loading approach whenever possible. Another noteworthy advantage of SMPs as drug delivery vehicles can be observed in the field of cardiovascular stents. Traditionally, these stents are composed of metals and require a polymeric coating with incorporated drugs for drug delivery purposes. This necessitates additional fabrication steps, leading to increased costs. In contrast, SMPs allow for direct incorporation of the drug within the matrix, thereby reducing fabrication costs.[42,43] Wiscke et al. have published a comprehensive review that provides a summary of the research conducted on drug-releasing SMPs.[44]

#### *2.5.1.2 CARDIOVASCULAR APPLICATIONS OF SHAPE-MEMORY POLYMERS*

SMPs have made notable advancements in the cardiovascular field, where they have found applications in the development of intravascular stents. In fact, several research groups have successfully developed intravascular stents using SMPs. Moreover, the self-expanding and biodegradable Remedy peripheral stent has even been commercialized, further highlighting the practical implementation of SMPs in cardiovascular interventions.[45] While the effectiveness of this stent has not reached the same level as self-expanding nitinol stents, this notable example demonstrates the tremendous potential of SMPs in seeking cardiovascular applications.[46] Stents based on SMPs exhibit significant potential for addressing intracranial stenosis. Unlike conventional stents made of stainless steel or shape-memory alloys, which are too rigid for delivery in neurovascular blood vessels, shape-memory polymer stents offer greater flexibility. This is particularly advantageous due to the small size and intricate twisting and turning of these blood vessels within the brain.[47] The considerably lower glassy modulus of SMPs in comparison to metals makes these materials more suitable for creating appropriate structures.[42]

Apart from intravascular stents, researchers have also explored other cardiovascular applications. Montgomery et al. have documented the development of a functional tissue patch utilizing poly(octamethylene maleate (anhydride) citrate). This innovative patch offers the potential for organ repair through a minimally invasive approach using injection. Remarkably, the patch can be injected through a mere 1 mm wide opening and subsequently expand in vivo to its original shape, without compromising cell viability. The researchers conducted successful in vivo testing of their cardiac patches in a rat model and achieved successful patch delivery to various target locations in pigs.[36] This approach offers a substitute for the conventional open-heart surgery typically required for similar injuries. The latter case serves as an illustration of how SMPs can significantly decrease the invasiveness of certain surgical procedures. Furthermore, SMPs show promise in the treatment of endovascular thromboses. For instance, Small et al. have described a shape-memory device that combines nitinol with a shape-memory polymer to create a device capable of removing blood clots in an in vitro setting.[48] The Maitland group also pursued a similar approach; however, they utilized a laser-activated device composed entirely of a shape-memory polymer.[49] While the examples provided are not exhaustive, they highlight the potential of SMPs in serving various cardiovascular applications. Numerous other SMPs hold promise for further advancements in the field.

#### *2.5.1.3 BONE TISSUE ENGINEERING USING SHAPE-MEMORY POLYMERS*

SMPs have demonstrated their potential in the repair of bone defects as well. An illustrative example is the work of Zhai et al., who developed composite scaffolds using poly(N-acryloyl glycinamide) and nano clay. These scaffolds effectively promoted osteogenic differentiation and stimulated the regeneration of bone tissue.[50] Furthermore, in a recent study by Tian et al., poly(ε-caprolactone) (PCL)-based networks were employed, incorporating hydroxyapatite. To achieve the integration of hydroxyapatite into the network, they modified it with thiol groups, allowing it to react with acrylate-modified PCL through a click reaction, resulting in a crosslinked network.[51] Wang et al. utilized 3D printing to create shape-memory scaffolds that incorporated superparamagnetic iron nanoparticles for osteogenic induction. They formulated their ink with a combination of gelatin and PEG, enhancing the printability of the scaffolds and resulting in improved fixity and recovery ratios. Notably, the scaffolds containing PEG exhibited enhanced osteogenesis capabilities.[34] One of the key advantages of SMPs over other materials used for bone defect repair is their ability to be heated to a malleable state, allowing them to conform to irregular bone defects. Upon cooling, the polymer retains its shape, effectively locking itself within the defect. Similarly, the repair of cartilage defects can also be achieved using SMPs. In a recent publication by Xuan et al., they explored the use of SMPs for cartilage regeneration. Their polymer formulation included bioactive ketogenic compounds, which facilitated the stimulation of cartilage regeneration. Once again, the main advantage of SMPs lies in their minimally invasive approach. They can be implanted in a temporary shape that is smaller in size than the final desired shape, making them better suited for limited space and the ability to accommodate irregular defects.[52]

#### *2.5.1.4 SHAPE-MEMORY POLYMERS WITH ANTIBACTERIAL FUNCTIONALITY*

The antibacterial functionality of SMPs is a valuable trait for all in vivo applications. This property significantly reduces the risk of infection, rendering the material more suitable for biomedical purposes. It is important to note that SMPs themselves do not possess inherent antibacterial properties. However, the incorporation of specific additives is necessary to confer the desired antibacterial effect.[36] The addition of additives serves a dual purpose for SMPs. Firstly, they enable the polymer to exhibit photoresponsivity, a characteristic not inherent in the virgin polymer. Secondly, these additives introduce antibacterial properties to the polymer. This combination presents an intriguing opportunity for researchers seeking to develop photoresponsive SMPs. When selecting additives to achieve photoresponsivity, considering the added benefit of antibacterial functionality becomes an important factor to consider.

## 2.6 POROUS ORGANIC POLYMERS FOR BIOMEDICAL APPLICATIONS

POPs, as a novel class of porous materials, have garnered significant attention and have already demonstrated potential applications in diverse fields such as gas adsorption/separation, heterogeneous catalysis, sensing, luminescence, and drug delivery. Recent advancements in the design and synthesis of new POPs have further expanded their application possibilities. To unlock novel applications, the priority lies in developing POPs with uncompromised physicochemical stability, well-controlled morphology, and reasonable processability. Introducing functional moieties into POPs opens up avenues for novel properties and applications. For instance, incorporating redox-active units enables energy storage, improving conductivity for electrochemical applications, or loading therapeutic drugs for cancer diagnosis and therapy. The design and preparation of new POPs for biomedical applications remain an active research area, driven by the unique structures and characteristics of POPs. These materials offer tremendous opportunities in the biomedical field, boasting design flexibility, high surface area, porosity, tunable skeleton structure, and ease of functionalization.

In recent years, POPs have witnessed widespread adoption in various biomedical applications, including drug delivery, encapsulation of biomacromolecules, photothermal and photodynamic therapy, biosensing, bioimaging, antibacterial applications, bioseparation, and many other emerging biomedical fields. This section aims to highlight the recent notable advancements in the utilization of POPs for diverse biomedical applications, aiming to inspire researchers to further explore and expand the use of POPs in the biomedical realm.

### *2.6.1 DRUG DELIVERY*

One of the biggest challenges in dealing with human health is the efficient delivery of drugs into the targeted position of our body in a controlled manner. Various types of materials have been developed as drug molecule carriers, such as natural and synthetic polymers,[53] hydrogels,[54] liposomes,[55] etc. An ideal drug carrier should possess the ability to efficiently and selectively deliver drugs while demonstrating high drug loading capacity and minimal cytotoxicity. However, many existing drug carriers exhibit limited drug loading capacities, emphasizing the need to explore more effective drug delivery systems. POPs, characterized by their high porosity, adjustable pore shape, and size, good biocompatibility, and customizable structure, represent promising candidates for drug delivery applications. Numerous POPs with high drug loading capacities have been reported in recent years, capable of controlled drug release either in response to external stimuli or through controlled release mechanisms. POPs are versatile carriers that can accommodate not only small-molecule pharmaceuticals but also biomacromolecule drugs such as proteins and nucleic acids. Strategies for loading drug molecules into POPs include physical entrapment within the pores, supramolecular interactions, intercalation between POP layers, and covalent bond attachment. Given that drug delivery systems commonly operate in physiological media, POPs with high water dispersibility and solubility are highly desirable for efficient drug delivery. One specific example of the potential of drug delivery using POPs can be seen in oral insulin delivery for the treatment of diabetes. Diabetes is a chronic disease and a leading cause of mortality worldwide, often necessitating insulin therapy. Traditional subcutaneous insulin injections are inconvenient, painful, and anxiety-inducing for patients. Therefore, developing an effective oral insulin delivery system is of paramount importance, offering a non-invasive and convenient therapeutic approach. An efficient oral insulin delivery system should exhibit biocompatibility, high insulin loading capacity, protection against the acidic stomach environment, and prevention of insulin degradation by enzymes. In this context, Benyettou et al. recently reported the use of imine-linked covalent organic framework (COF) nanoparticles for efficient in vivo oral insulin delivery, demonstrating promising advances in the field.[56]

### 2.6.2 BIOMACROMOLECULE IMMOBILIZATION

Biomacromolecules, such as enzymes and DNA, play a crucial role in various biomedical applications, including biocatalysis, gene therapy, and cancer treatment. However, their inherent instability and susceptibility to degradation under ex vivo conditions pose significant challenges. While enzymes exhibit remarkable biocatalytic efficiency, their structural fragility hinders their recycling and repeated usage. To overcome these limitations, immobilizing these biomacromolecules within suitable hosts has emerged as a promising strategy to safeguard them from decomposition. Enzymes, being vital biomacromolecules in life, serve as indispensable biocatalysts for numerous biological processes.[57] The encapsulation of enzymes within porous materials has garnered significant interest due to its potential applications in the biomedical and pharmaceutical industries. Enzymes can be immobilized in POPs through physical adsorption or covalent bond attachment. In 2015, Banerjee's group successfully synthesized a hollow COF and utilized it as a host for enzyme immobilization.[58] Due to the relatively large size of enzyme molecules, it is essential for the pore apertures of POPs to be sufficiently spacious to accommodate enzymes. Consequently, the development of POPs with mesopores is highly sought after for effective enzyme immobilization.

### 2.6.3 PHOTOTHERMAL AND PHOTODYNAMIC THERAPY

Photothermal therapy (PTT) and photodynamic therapy (PDT) represent two extensively employed and highly effective photophysical therapeutic approaches for the eradication of cancer cells.[59] PTT utilizes photothermal agents to absorb light and transform it into heat energy, leading to an increase in temperature that selectively targets and eliminates abnormal cells.[60] In PDT, photosensitizers absorb light and excite oxygen, initiating a cascade of photochemical reactions that generate reactive oxygen species (ROS). These activated ROS possess potent oxidative properties, causing DNA injury and protein damage, resulting in cellular harm, necrosis, and eventual demise.[61] PTT and PDT employ distinct mechanisms for eliminating cancer cells, yet both involve the absorption of light. In order to efficiently absorb light, molecules or materials utilized in PTT or PDT typically possess a low band gap. POPs offer numerous advantages for both PTT and PDT, including high porosity, tunable structure, and the potential for a large p-conjugation system. On one hand, POPs can be modularly designed and synthesized through molecular engineering, enabling them to absorb a substantial amount of light across the visible to near-infrared (NIR) spectrum. On the other hand, the internal voids or unreacted functional groups of POPs facilitate the incorporation of photosensitizers or photothermal agents to create active PTT/PDT POPs. Over the past few years, numerous POPs have been reported for PTT and PDT cancer treatment. Furthermore, the combination of PTT and PDT for cancer treatment has been extensively studied due to its enhanced synergistic ability to kill cancer cells.

In the realm of biosensing, POPs have emerged as versatile platforms for detecting a wide range of biomolecules, including glucose, peptides, proteins, and DNA. POPs offer potential advantages such as minimal or non-toxicity, biocompatibility, and purely organic characteristics. The sensing mechanism primarily relies on interactions such as hydrogen bonding and p-p interactions between the POPs and the target molecules. For instance, in 2014, Fang et al. reported an imine-linked COF with abundant amino groups capable of absorbing proteins and DNA, demonstrating the strong interaction between the COF backbone and biomolecules. Subsequently, more research has focused on utilizing POPs as biosensing platforms.[62]

### *2.6.4 BIOIMAGING*

Bioimaging technology plays a crucial role in the early diagnosis and treatment of numerous diseases, enabling non-invasive visualization of biological processes in real time with minimal interference.[63] The prevailing method for bioimaging involves generating targeted photoluminescence, enabling precise identification of affected cells and tissues. Prominent bioimaging technologies encompass fluorescence imaging and magnetic resonance imaging.

### *2.6.5 ANTIBACTERIAL*

Bacterial infections pose a significant risk to human health, often resulting in severe diseases and even death. Consequently, the urgent need to combat this threat has underscored the importance of developing exceptionally potent antibacterial materials. Numerous material variants have been devised to actively inhibit or minimize the proliferation of harmful bacteria.[64] In recent years, polymeric organic frameworks (POFs) have emerged as highly promising contenders for antibacterial applications due to their adaptable structural design and seamless integration of antibacterial agents. These agents encompass a wide range of compounds, such as quaternary ammonium cations, silver chloride (AgCl), calamine, and other positively charged functional groups. Among these, quaternary pyridinium ions and their derivatives have gained recognition as notable antibacterial moieties.[65]

### *2.6.6 BIOSEPARATION*

Bioseparation is a critical process involving the separation of various biological substances, including biomacromolecules, cells, and cell components. This process holds immense significance in numerous biotechnological and biopharmaceutical applications. Additionally, POFs have garnered significant attention for their extensive utilization in gas/ liquid separation.[66] Given their ability to regulate pore size and customize functionality, the utilization of POFs holds immense potential for biomacromolecule separation. By strategically adjusting precursor materials and incorporating specific functional groups, POFs with tailored pore sizes and functionalities can be synthesized. Regarding bioseparation, mechanisms such as size selectivity and charge repulsion come into play. The progress of POFs in biomedical applications and beyond greatly depends on interdisciplinary collaboration. The concerted efforts of scientists and engineers from diverse backgrounds will expedite the translation of POFs from research laboratories to clinical trials. Chemists and materials scientists excel in designing and fabricating various POFs with versatile functionalities while biomedical engineers and clinicians contribute by evaluating their biomedical applications from a clinical perspective. Although challenges remain, substantial achievements in biomedical applications have paved the way for the versatile nature of POFs. We firmly believe that through collaborative multidisciplinary endeavors, there will be a wealth of exciting results utilizing POFs in biomedical applications, ultimately leading to their clinical implementation in the near future.

## 2.7 NATURAL BASE BIOPOLYMERS FOR BIOMEDICAL APPLICATIONS

Natural base biopolymers have gained significant attention in the field of biomedical applications due to their inherent biocompatibility, biodegradability, and structural similarity to biological tissues. Here are some commonly used natural base biopolymers for biomedical applications:

Collagen: Collagen is the most abundant protein in the human body and provides structural support to various tissues. It is widely used in tissue engineering, wound healing, and drug delivery applications due to its biocompatibility and cell-adhesive properties.

Chitosan: Chitosan is derived from chitin, a natural polysaccharide found in the exoskeletons of crustaceans. It has excellent biocompatibility, biodegradability, and antimicrobial properties, and can be easily processed into various forms such as films, gels, and sponges. Chitosan is used in wound dressings, drug delivery systems, and tissue engineering scaffolds.

Hyaluronic Acid (HA): HA is a glycosaminoglycan naturally found in the extracellular matrix of various tissues. It has excellent water-holding capacity and lubricating properties. HA is used in ophthalmology, orthopedics, and tissue engineering for applications such as viscoelastic supplements, drug delivery systems, and scaffolds.

Gelatin: Gelatin is derived from the hydrolysis of collagen and has similar properties, such as biocompatibility and cell-adhesive properties. It is widely used in pharmaceuticals, wound dressings, and tissue engineering scaffolds.

Alginate: Alginate is extracted from brown seaweeds and has excellent biocompatibility and gel-forming properties. It is commonly used in drug delivery systems, wound dressings, and tissue engineering due to its ability to form hydrogels in the presence of divalent cations.

Silk Fibroin: Silk fibroin is derived from silkworm cocoons and has been used for centuries in textile manufacturing. It has good mechanical properties, biocompatibility, and biodegradability. Silk fibroin is used in tissue engineering, wound healing, and drug delivery systems.

Fibrin: Fibrin is a protein involved in the blood clotting process. It can be obtained from human or animal blood and has been used as a natural scaffold for tissue engineering, wound healing, and drug delivery applications.

These natural base biopolymers offer a wide range of properties and can be tailored through modifications to suit specific biomedical applications. Researchers continue to explore and develop new strategies to enhance their functionality and expand their potential applications in the biomedical field.

## 2.8 CONCLUSION

In conclusion, the versatile nature of polymers has propelled them to the forefront of biomedical research and application, offering innovative solutions across various domains. In hard tissue applications, polymers have demonstrated remarkable potential in orthopedics and dentistry, enabling the development of biocompatible implants, scaffolds, and controlled drug delivery systems. Their adaptability and ability to mimic natural tissues hold great promise for enhancing patient outcomes in these fields. Soft tissue applications have harnessed the regenerative properties of polymers, fostering wound healing, tissue repair, and controlled drug release. This holds significant potential for addressing chronic wounds and promoting faster recovery.

The exploration of SMPs has ushered in a new era of responsive and adaptable biomedical devices, facilitating precise drug release and minimally invasive interventions. POPs have expanded the possibilities of drug delivery and bioseparation, allowing for targeted therapy and personalized medicine approaches.

Furthermore, the emergence of natural base biopolymers has revolutionized drug delivery and tissue engineering, emphasizing sustainability and biodegradability. As we move forward, continued research and development in these diverse areas of polymer-based biomedical applications are expected to lead to groundbreaking advancements in healthcare. The integration of polymers into clinical practice has the potential to enhance patient care, improve treatment outcomes, and contribute to the overall well-being of individuals worldwide.

## KEYWORDS

- **biopolymer**
- **biomedical**
- **bioimaging**
- **scaffolds**
- **drug delivery**
- **healthcare**

## REFERENCES

1. Francis, A. Biological Evaluation of Preceramic Organosilicon Polymers for Various Healthcare and Biomedical Engineering Applications: A Review. *J. Biomed. Mater. Res. Part B Appl. Biomater.* **2021,** *109*, 744–764. https://doi.org/10.1002/jbm.b.34740
2. Tibbitt, M. W.; Rodell, C. B.; Burdick, J. A.; Anseth, K. S. Progress in Material Design for Biomedical Applications. *Proc. Natl. Acad. Sci.* **2015,** *112*, 14444–14451. https://doi.org/10.1073/pnas.1516247112
3. Cheung, H.-Y.; Lau, K.-T.; Lu, T.-P.; Hui, D. A Critical Review on Polymer-Based Bio-Engineered Materials for Scaffold Development. *Compos. Part B Eng.* **2007,** *38*, 291–300. https://doi.org/10.1016/j.compositesb.2006.06.014
4. Turner, B. L.; Senevirathne, S.; Kilgour, K. et al. Ultrasound-Powered Implants: A Critical Review of Piezoelectric Material Selection and Applications. *Adv. Healthc. Mater.* **2021,** *10*. https://doi.org/10.1002/adhm.202100986
5. Sharma, R.; Malviya, R.; Singh, S.; Prajapati B A Critical Review on Classified Excipient Sodium-Alginate-Based Hydrogels: Modification, Characterization, and Application in Soft Tissue Engineering. *Gels* **2023,** *9*, 430. https://doi.org/10.3390/gels9050430
6. Wancket, L. M.; Schuh, J. C. L.; Drevon-Gaillot, E. *Biomedical Materials and Devices*; Elsevier, 2023.
7. Cecen, B.; Hassan, S.; Li, X.; Zhang, Y. S. Smart Biomaterials in Biomedical Applications: Current Advances and Possible Future Directions. *Macromol Biosci.* 2023. https://doi.org/10.1002/mabi.202200550
8. Wang, K.; Man, L.; Zhang, M. et al. Programmable Polymers with Shape Memory for Biomedical Applications. *Program Mater.* **2023,** *1*, e2. https://doi.org/10.1017/pma.2023.2
9. Zhao, L.; Zhang, J.; Liang, H. et al. Natural Polymer-Based Hydrogels: From Polymer to Biomedical Application. **2023,** 1–24. https://doi.org/10.20944/preprints202309.0901.v1
10. Elango, J.; Bao, B.; Wu W Development of 3D-Printed Biocompatible Materials for Tendons Substitution. In: *Cartilage Tissue and Knee Joint Biomechanics*; Elsevier, 2024; pp 453–465.
11. Bakhshi, R.; Mohammadi-Zerankeshi, M.; Mehrabi-Dehdezi, M. et al. Additive Manufacturing of PLA-Mg Composite Scaffolds for Hard Tissue Engineering Applications. *J. Mech. Behav. Biomed. Mater.* **2023,** *138*, 105655. https://doi.org/ 10.1016/j.jmbbm.2023.105655
12. Sun, X.; Heng, B. C.; Zhang, X Oral Hard Tissue Defect Models for Evaluating the Regenerative Efficacy of Implant Materials. *MedComm—Biomater. Appl.* **2023,** *2*. https://doi.org/10.1002/mba2.38
13. Arpit, S.; Jaya, T. *Properties and Applications of Superabsorbent Polymers*; Springer Nature Singapore: Singapore.
14. Zhang, X.; Malekahmadi, O.; Mohammad Sajadi, S. et al. Thermomechanical Properties of Coated PLA-3D-Printed Orthopedic Plate with PCL/Akermanite Nano-Fibers: Experimental Procedure and AI Optimization. *J. Mater. Res. Technol.* **2023.** https://doi.org/10.1016/j.jmrt.2023.09.215

15. Sivakumar, P. M.; Yetisgin, A. A.; Demir, E. et al. Polysaccharide-Bioceramic Composites for Bone Tissue Engineering: A Review. *Int. J. Biol. Macromol.* **2023,** *250*, 126237. https://doi.org/10.1016/j.ijbiomac.2023.126237
16. Liang, W.; Zhou, C.; Zhang, H. et al. Recent Advances in 3D Printing of Biodegradable Metals for Orthopaedic Applications. *J. Biol. Eng.* **2023,** *17*, 1–24. https://doi.org/10.1186/s13036-023-00371-7
17. Monia, T.; Ridha, B. C. Polymer-Ceramic Composites for Bone Challenging Applications: Materials and Manufacturing Processes. *J. Thermoplast Compos. Mater*. **2023**. https://doi.org/10.1177/08927057231190066
18. Bańkosz, M.; Urbaniak, M. M.; Szwed, A. et al. Physicochemical and Biological Analysis of Composite Biomaterials Containing Hydroxyapatite for Biological Applications. *J. Biomed. Mater. Res. Part B Appl. Biomater.* **2023**. https://doi.org/10.1002/jbm.b.35309
19. Atia, G. A.; Shalaby, H. K.; Roomi, A. B. et al. Polymeric Biomaterials in Therapeutic, and Regenerative Orofacial Applications Macro, Micro, and Nano-Inspired Bioactive Polymeric Biomaterials in Therapeutic, and Regenerative Orofacial Applications, 2023.
20. Chen, Y.; Wang, X.; Tao, S. et al. Research Advances in Smart Responsive-Hydrogel Dressings with Potential Clinical Diabetic Wound Healing Properties. *Mil. Med. Res.* **2023,** *10*, 37. https://doi.org/10.1186/s40779-023-00473-9
21. Mensah, A.; Rodgers, A. M.; Larrañeta, E. et al. Treatment of Periodontal Infections, the Possible Role of Hydrogels as Antibiotic Drug-Delivery Systems. *Antibiotics* **2023,** *12*, 1073. https://doi.org/10.3390/antibiotics12061073
22. Huang, B.; Yang, M.; Kou, Y.; Jiang B Absorbable Implants in Sport Medicine and Arthroscopic Surgery: A Narrative Review of Recent Development. *Bioact. Mater.* **2024,** *31*, 272–283. https://doi.org/10.1016/j.bioactmat.2023.08.015
23. Zhang, Y.; Zhang, C.; Li, Y. et al. Evolution of Biomimetic ECM Scaffolds from Decellularized Tissue Matrix for Tissue Engineering: A Comprehensive Review. *Int. J. Biol. Macromol.* **2023,** *246*, 125672. https://doi.org/10.1016/j.ijbiomac.2023.125672
24. Wankar, J.; Kotla, N. G.; Gera, S. et al. Recent Advances in Host–Guest Self-Assembled Cyclodextrin Carriers: Implications for Responsive Drug Delivery and Biomedical Engineering. *Adv. Funct. Mater.* 2020, *30*. https://doi.org/10.1002/adfm.201909049
25. Xiao, R.; Huang, W. M. Heating/Solvent Responsive Shape-Memory Polymers for Implant Biomedical Devices in Minimally Invasive Surgery: Current Status and Challenge. *Macromol. Biosci.* **2020,** *20*. https://doi.org/10.1002/mabi.202000108
26. Husain, I.; Luo, X Novel Approaches to Immunomodulation for Solid Organ Transplantation. *Annu. Rev. Med.* **2024,** *75*. https://doi.org/10.1146/annurev-med-050522-034012
27. Tiwari, G.; Tiwari, R.; Bannerjee, S. et al. Drug Delivery Systems: An Updated Review. *Int. J. Pharm. Investig.* **2012,** *2*, 2. https://doi.org/10.4103/2230-973X.96920
28. Chen, M.-C.; Tsai, H.-W.; Chang, Y.; Lai, W.-Y.; Mi, F.-L.; Liu, C.-T.; Wong, H.-S.; H.W. Sung Rapidly Self-Expandable Polymeric Stents with a Shape-Memory Property. *Biomacromolecules* **2007,** *8*, 2774–2780.
29. Correia, C. O.; Caridade, S. G. et al. Chitosan Membranes Exhibiting Shape Memory Capability by the Action of Controlled Hydration. *Polymers* **2014,** *6*, 1178–1186.

30. Li, S.-t.; Jin, X.-z.; Shao, Y.-w.; Qi, X.-d.; Yang, J. et al. Gold Nanoparticle/Reduced Graphene Oxide Hybrids for Fast Light-Actuated Shape Memory Polymers with Enhanced Photothermal Conversion and Mechanical Stiffness. *Eur. Polym. J.* **2019,** *116*, 302–310.
31. Xie, H.; Shao, J.; Ma, Y.; Wang, J.; Huang, H.; Yang, N.; Wang, H.; Ruan, C.; Luo, Y.; Wang, Q. Q.; Chu, P. K. Biodegradable Near-Infrared-Photoresponsive Shape Memory Implants Based on Black Phosphorus Nanofillers. **2018,** *164*, 11–21.
32. Kim, B. K.; Lee, S. Y. et al. Polyurethanes Having Shape Memory Effects. *Polymer* **1996,** *37*, 5781–5793.
33. Alavijeh, R. Z.; Shokrollahi, P. et al. A Thermally and Water Activated Shape Memory Gelatin Physical Hydrogel, with a Gel Point Above the Physiological Temperature, for Biomedical Applications. *J. Mater. Chem.* **2017,** *5*, 2302–2314.
34. Wang, C.; Wang, H.; Zou, F.; Chen, S. et al. Development of Polyhydroxyalkanoate-Based Polyurethane with Water-Thermal Response Shape-Memory Behavior as New 3D Elastomers Scaffolds. *Polymers* **2019,** *11* (6), 1030.
35. Zini, E.; Scandola, M.; Dobrzynski, P.; Kasperczyk, J. et al. Shape Memory Behavior of Novel (L-Lactide–Glycolide–Trimethylene Carbonate) Terpolymers. *Biomacromolecules* **2007,** *8*, 3661–3667.
36. Ahadian, S.; Huyer, L. D.; Rito, M. L.; Civitarese, R. A.; Vanderlaan, R. D. et al. Flexible Shape-Memory Scaffold for Minimally Invasive Delivery of Functional Tissues. *Nat. Mater.* **2017,** *16*, 1038–1046.
37. Chen, H.-M.; Wang, L. et al. Recent progress in Shape Memory Polymers for Biomedical Applications. *Chinese J. Polym. Sci.* **2018,** *36*, 905–917.
38. Wang, K.; Strandman, S. et al. A Mini Review: Shape Memory Polymers for Biomedical Applications. *Front. Chem. Sci. Eng.* **2017,** 143–153.
39. Wischke, C.; Neffe, A. T.; Steuer, S.; Engelhardt, E. et al. AB-Polymer Networks with Cooligoester and Poly(n-butyl acrylate) Segments as a Multifunctional Matrix for Controlled Drug Release. *Macromol. Biosci.* **2010,** *9*, 1063–1072.
40. Wischke, C.; Neffe, A. T.; Steuer, S. et al. Evaluation of a Degradable Shape-Memory Polymer Network as Matrix for Controlled Drug Release. *J. Control Release* **2009,** *3*, 243–250.
41. Jahangiri, M.; Kalajahi, A. E.; Rezaei, M. et al. Shape Memory Hydroxypropyl Cellulose-g-poly (ε-caprolactone) Networks with Controlled Drug Release Capabilities. *J. Polym. Res.* **2019,** 1–14.
42. Small, T. S. W.; Benett, W. J.; Matthews, D. L.; Hartman, J. et al. Fabrication and In Vitro Deployment of a Laser-Activated Shape Memory Polymer Vascular Stent. *Biomed. Eng. Online* **2007,** 1–8.
43. Wischke, C.; Behl, M. et al. Drug-Releasing Shape-Memory Polymers—The Role of Morphology, Processing Effects, and Matrix Degradation. *Expert Opin. Drug Deliv.* **2013,** *9*, 1193–1205.
44. Bontinck, J.; Goverde, P.; Schroë, H.; Hendriks, J.; Maene, L. et al. Treatment of the Femoropopliteal Artery with the Bioresorbable Remedy Stent. *J. Vasc. Surg.* **2016,** *5*, 1311–1319.
45. Jia, H.; Gu, S.-Y. et al. 3D Printed Self-Expandable Vascular Stents from Biodegradable Shape Memory Polymer. *Adv. Polym. Technol.* **2018,** *37*, 3222–3228.

46. Tan, L.; Hu, J.; Huang, H.; Han, J. et al. Study of Multi-Functional Electrospun Composite Nanofibrous Mats for Smart Wound Healing. *Int. J. Biol. Macromol.* **2015,** 469–476.
47. Baer, G. M.; Wilson, T. S.; Small, W.; Hartman, J.; Benett, W. J.; Matthews, D. L.; Maitland, D. J. Thermomechanical Properties, Collapse Pressure, and Expansion of Shape Memory Polymer Neurovascular Stent Prototypes. *J. Biomed. Mater. Res. Part B Appl. Biomater.* **2009,** *90*, 421–429.
48. Small, W.; Wilson, T. S.; Buckley, P. R.; Benett, W. J.; Loge, J. M.; Hartman, J. et al. Prototype Fabrication and Preliminary In Vitro Testing of a Shape Memory Endovascular Thrombectomy Device. *IEEE Trans. Biomed. Eng.* **2007,** *54*, 1657–1666.
49. Behl, M.; Zotzmann, J. et al. Shape-Memory Polymers and Shape-Changing Polymers. *Shape-Memory Polym.* **2009,** 1–40.
50. Xuan, H.; Hu, H.; Geng, C.; Song, J.; Shen, Y.; Lei, D.; Guan, Q.; Zhao, S. et al. Biofunctionalized Chondrogenic Shape-Memory Ternary Scaffolds for Efficient Cell-Free Cartilage Regeneration. *Acta Biomater*, **2020,** 97–110.
51. Zhang, Y.; Zhou, S.; Chong, K. C.; Wang, S. et al. Near-Infrared Light-Induced Shape Memory, Self-Healable and Anti-Bacterial Elastomers Prepared by Incorporation of a Diketopyrrolopyrrole-Based Conjugated Polymer. *Mater. Chem. Front.* **2019,** 836–841.
52. Tian, G.; Zhu, G.; Xu, S. et al. A Novel Shape Memory Poly(ε-caprolactone)/Hydroxyapatite Nanoparticle Networks for Potential Biomedical Applications. *J. Solid State Chem.* **2019,** *272*, 78–86.
53. Zhou, M.; Zhang, X.; Xie, J.; Qi, R.; Lu, H.; Leporatti, S.; Chen, J. et al. pH-Sensitive Poly(β-amino ester) s Nanocarriers Facilitate the Inhibition of Drug Resistance in Breast Cancer Cells. *Nanomaterials* **2018,** *8*, 952.
54. Lei, Z.; Zhu, W.; Zhang, X.; Wang, X. et al. Bio-Inspired Ionic Skin for Theranostics. *Adv. Funct. Mater.* **2021,** *31*.
55. Ouyang, J.; Ji, X.; Zhang, X.; Feng, C.; Tang, Z.; Kong, N. et al. In Situ Sprayed NIR-Responsive, Analgesic Black Phosphorus-Based Gel for Diabetic Ulcer Treatment. *Proc. Natl. Acad. Sci.* **2020,** *46*, 28667–28677.
56. Kaddour, T. P.; Das, G.; Sharma, S. K.; Thomas, S. A. et al. In Vivo Oral Insulin Delivery via Covalent Organic Frameworks. *Chem. Sci.* **2021,** *12*, 6037–6047.
57. Gao, D.; Chen, T.; Chen, S.; Ren, X.; Han, Y.; Li, Y. et al. Targeting Hypoxic Tumors with Hybrid Nanobullets for Oxygen-Independent Synergistic Photothermal and Thermodynamic Therapy. *Nano-Micro Lett.* **2021,** *13*, 1–21.
58. Liang, S.; Wu, X.-L.; Xiong, J.; Zong, M.-H. et al. Metal-Organic Frameworks as Novel Matrices for Efficient Enzyme Immobilization: An Update Review. *Coord. Chem. Rev.* **2020,** 213149.
59. Ji, X.; Ge, L.; Liu, C.; Tang, Z.; Xiao, Y.; Chen, W. et al. Capturing Functional Two-Dimensional Nanosheets from Sandwich-Structure Vermiculite for Cancer Theranostics. *Nat. Commun.* **2021,** *12*.
60. Yang, Z.; Gao, D.; Guo, X.; Jin, L.; Zheng, J.; Wang, Y. et al. Fighting Immune Cold and Reprogramming Immunosuppressive Tumor Microenvironment with Red Blood Cell Membrane-Camouflaged Nanobullets. *ACS Nano* **2020,** *14*, 17442–17457.
61. Li, X.; Lee, S.; Yoon, J. Supramolecular Photosensitizers Rejuvenate Photodynamic Therapy. *Chem. Soc. Rev.* **2018,** 1174–1188.

62. Yang, R.; Gong, Z.; Zhang, X. et al. Single-Walled Carbon Nanotubes (SWCNTs) and Poly (3, 4-ethylenedioxythiophene) Nanocomposite Microwire-Based Electronic Biosensor Fabricated by Microlithography and Layer-by-Layer Nanoassembly. *J. Nanosci. Nanotechnol.* **2019,** *19*, 7591–7595.
63. Wallyn, J.; Anton, N.; Akram, S. et al. Biomedical Imaging: Principles, Technologies, Clinical Aspects, Contrast Agents, Limitations and Future Trends in Nanomedicines. *Pharm. Res.* **2019,** *36*, 1–31.
64. Huang, R.; Hu, M.; Liang, W.; Zheng, J.; Du, Y.; Lin, Y. et al. One-Step Preparation of Green Fabric for Continuous Antibacterial Applications. *Engineering* **2021,** *7*, 326–333.
65. Zhao, T.; Sun, G. Hydrophobicity and Antimicrobial Activities of Quaternary Pyridinium Salts. *J. Appl. Microbiol.* **2008,** *104*, 824–830.
66. Du, Y.; Huang, Z.; Wu, S.; Xiong, K.; Zhang, X.; Zheng, B.; Nadimicherla, R.; Fu, R. et al. Preparation of Versatile Yolk-Shell Nanoparticles with a Precious Metal Yolk and a Microporous Polymer Shell for High-Performance Catalysts and Antibacterial Agents. *Polymer* **2018,** *137*, 195–200.

# CHAPTER 3

# Sustainable Polymer Nanocomposites for Energy Applications

NIVYA ROY[1], JIJI ABRAHAM[2], SRILAKSHMI P. BHASKER[2], SUSAN SAMUEL[2], NIMITHA K. C.[2,3], EUNICA K. SUNNY[2], and SALIYA C. S.[2]

[1]*Carmel College (Autonomous), Mala, Thrissur, Kerala, India*

[2]*Vimala College (Autonomous), Thrissur, Kerala, India*

[3]*Christ College (Autonomous), Irinjalakuda, Kerala, India*

## ABSTRACT

Polymer nanocomposites (PNCs) exhibit unique physiochemical characteristics because of the synergistic combination of properties of filler and polymer. Studies have demonstrated that PNCs can be used for energy conversion and energy storage of energy due to their flexibility, lightweight nature, high chemical resistance, good processability, eco-fiendly, and acceptable energy performance. This chapter focuses on the recent advancements in applications of PNCs such as thermoelectric generators (TEGs), photovoltaic cells, supercapacitors, fuel cell devices, and hydrogen storage devices. Later, energy applications of piezoelectric PNCs are reviewed with special emphasis on energy storage, energy harvesting, and storage applications. Finally, a summary and future outlook are presented to highlight future directions of research in this emerging and vital sector.

Polymers and Functional Materials for a Cleaner Environment and Human Health: Sustainable Technologies and New Research. Ruby Varghese, Ann Rose Abraham, Shrikaant Kulkarni, & A. K. Haghi (Eds.)

## 3.1 AN INTRODUCTION TO POLYMER NANOCOMPOSITES

Polymers have a wide variety of applications because of their unique traits: lightweight, ease of production, and often ductile nature. However, there are some drawbacks to polymers. To improve their overall performance, polymers are commonly reinforced with nanosized fillers to improve different properties and hence open up new novel applications. The nanoscale ranges in the fillers are basic, from isotropic to extensive anisotropic sheet-like or needle-like morphologies, with at least one characteristic length scale of the order of nanometers. Thus, the combination of nanotechnology and polymer materials created a revolutionary change in the technological advancements. Polymer nanocomposites (PNCs) are classified into different categories based on the dimension of nanofillers. The first category is PNCs based on one-dimensional (1D) filler such as carbon nanotubes (CNTs), nanofibers, nanotubes, and halloysite nanotubes. The second category includes nanocomposites (NCs) based on two-dimensional (2D) fillers such as graphene, MXene, clay, layered silicates, etc. The third type of PNCs includes zero-dimensional (0D) nanofillers such as semiconductor nanoclusters, spherical silica, and quantum dots.[1] The uniform dispersion of nanofillers in the polymer matrix and the good interface between will and polymer are the two important prerequisites for achieving excellent properties. The ncorporation of nanofillers improves the overall characteristics of polymers, such as mechanical properties, rheological properties, viscoelastic properties, electric and electronic properties, catalytic properties, thermal properties, biodegradable properties, barrier properties, etc., at very low concentrations of fillers. Here are some of the applications of PNCs: automotive industry, aerospace industry, electronics and electrical industry, energy storage, coatings and paints, biomedical applications, packaging industry, construction industry, textiles and clothing, environmental applications, etc.

### *3.1.1 SYNTHESIS OF POLYMER NANOCOMPOSITES*

Fabrication routes of PNCs can be classified into the following categories:

- **First category: Melt intercalation**—It is the typical standard approach for fabricating thermoplastic PNCs. It entails adding the filler, annealing the polymer matrix at high temperatures, and then kneading the composite to achieve even distribution. Advantages of this technique include its environmentally friendly nature due to the lack of solvent usage and its economically feasible nature because of its compatibility with industrial processes such as injection molding and extrusion. Even though there are so many advantages, the high temperature used in this process can damage the surface functionalization of many fillers.[2]

- **Second category: In situ polymerization**—In situ polymerization refers to a polymerization process where the polymer is formed directly within the reaction mixture or in the presence of the material to be polymerized. Polymerization starts either using heat, radiation, initiator diffusion, or by organic initiator or catalyst fixed through cationic exchange. In between the interlayers, the monomers then polymerize to create intercalated or exfoliated NCs. This method has an advantage over melt and exfoliation adsorption methods in that it achieves better exfoliation.[3]
- **Third category: The solvent casting**—This technique involves the dispersion of nanoparticles (NPs) within a polymer matrix by using a solvent. By using vigorous agitation, such as magnetic stirring, high shear mixing, reflux, or sonication, the NP/polymer/solvent solution is blended. The polymer solution is poured into a mold once the NPs have been evenly scattered. After the solvent has completely evaporated, the composite film or sheet is left behind. Solution processing is frequently used to create NCs with film or sheet-like structure.[4]

### 3.1.2 *CHARACTERIZATION OF POLYMER NANOCOMPOSITES*

- **Spectroscopic techniques:** These techniques bring information about PNCs at the molecular level.
  Spectroscopic techniques for characterizing PNCs include the following:
  - Fluorescence spectroscopy,
  - Fourier transform infrared spectroscopy,
  - Raman spectroscopy,
  - Ultraviolet–visible (UV–Vis) spectroscopy,
  - Solid-state NMR Spectroscopy.[5]

- **Morphological characterization:** The morphological characterization of PNCs involves studying the arrangement, dispersion, and distribution of the NPs within the polymer matrix. Various techniques include, transmission electron microscopy (TEM), scanning electron microscopy (SEM), and atomic force microscopy (AFM).[6]
- **Mechanical characterization:** Mechanical characterization of PNCs involves assessing their mechanical properties to better realize their performance and considering the behavior. Some techniques include tensile testing, flexural testing, impact testing, hardness testing, rheological characterization, dynamic mechanical analysis (DMA), etc.[7]
- **Dielectric characterization:** Dielectric data could offer insightful details about the interactions between polymers and nanofillers, the distribution of nanoinclusions, and the morphology of NC materials, enhancing our understanding of the relationship between their structure and properties. Direct current (dc) conductivity tests and broadband dielectric spectroscopy (BDS) can be used to experimentally study the electrical behavior of NCs.[8]
- **Thermal characterization:** It involves understanding how the addition of NPs or nanofillers affects the thermal behavior of the polymer matrix. Various thermal characterization techniques include the following:
  - Differential scanning calorimetry (DSC),
  - Thermogravimetric analysis (TGA),
  - Thermal conductivity measurement,
  - Thermomechanical analysis (TMA), etc.[9]

## 3.2 APPLICATION OF POLYMER NANOCOMPOSITES FOR ENERGY STORAGE AND CONVERSION

As the population increases on daily basis, there is an exponential increase in the energy demand. Studies have demonstrated that PNCs can be used for energy conversion and energy storage applications because of their flexibility, lightweight nature, high chemical resistance, good processability, environment friendliness as well as improved energy performance.[10] In this chapter, applications of PNCs in various energy-related fields, such as thermoelectric generators (TEGs), solar photovoltaic cells, supercapacitors and batteries, fuel cells, and hydrogen storage, are discussed concisely. At the end of the chapter, piezoelectric and green polymeric NCs for energy applications are briefly discussed.

### 3.2.1 THERMOELECTRIC GENERATORS

One of the readily accessible energies is thermal energy, which is used in the different industrial sectors such as the operation of electronic devices, the operation of cars and indoor structures, etc. The Seebeck effect is applied successfully by TEGs to transform thermal energy directly into electrical energy. Additionally, they operate silently due to the lack of mechanical structures and/or dynamic parts. They are environmentally beneficial since they do not include chemical compounds that could be produced from different types of substrates, such as silicon, ceramics, and polymers. The emergence of conducting organic polymers and compound semiconductor nanostructures has increased interest in and research into polymer–inorganic composites for thermoelectric (TE) applications. Polymer–inorganic composites have been created by physical mixing, solution processing, and in situ synthesis of polymers and/or inorganic crystals, showing TE advantages over the separate constituents and producing meaningful power production and high practical efficiencies. While the polymer matrix still has a very poor thermal conductivity, the addition of conducting inorganic compounds in conjugated polymers is anticipated to raise composite electrical conductivity and can also improve the Seebeck coefficient. The TEGs' other benefits include dependability, a low maintenance requirement, and environmental safety. In a variety of industries, including the medical, industrial, and military, they can be used as a power source.[11]

Conductive polymers are commonly used in the construction of organic TE materials. The typical polymers are electrically insulating and cannot be employed as TE materials. Conjugated π-bonds are created when all of the carbon atoms at the backbones are $sp^2$ hybridized, allowing $2p$ electrons to become delocalized and move freely between atoms. The conductive polymers have high electrical conductivity by nature due to the conjugated π-bonds, which makes them desirable for practical TE applications. Along with conductive polymers, extensive research on carbon nanotubes (CNTs), the creation of composite materials, and fresh approaches to structural engineering all help to enhance the TE performance of polymer-based composites.

#### 3.2.1.1 CARBON NANOTUBES/CONDUCTIVE POLYMER COMPOSITES

CNT is necessary to improve the electrical conductivity of materials made of polymers because it is a conductive filler. To enhance the TE performance of polymers, both single- and double-walled carbon nanotubes (SWCNTs/DWCNTs) and multiwalled carbon nanotubes (MWCNTs) are utilized. However, the outcomes demonstrate that the electrical conductivity and Seebeck coefficient of the polymer-based composites benefit more from the S/DWCNTs. The Fermi level's density of the state is correlated with the Seebeck coefficient. The Seebeck coefficient rises sharply at a tiny Fermi energy shift away from the center of the bandgap of semiconducting-SWCNT (s-SWCNT), but beyond that, it falls as the Fermi level shift increases in size. However, the metallic SWCNTs (m-SWCNTs) have a nearly constant Fermi energy. The concentration of s-SWCNT can, therefore, have an impact on the Seebeck coefficient of SWCNTs networks. Additionally, a smaller diameter of SWCNTs and SWCNT bundles is advantageous to increase PF. The Seebeck coefficient and electrical conductivity in CNT-based composites are proportional to the CNT concentration. As a conductive filler, CNT content often correlates with a composite's electrical conductivity.[12]

#### 3.2.1.2 OTHER POLYMER NANOCOMPOSITES

Studies of polyaniline (PANI) show that PANI/nickel oxide (NiO) NCs are a promising candidate for TEG. Due to the ordered chain orientation and strong conjugation of the polymer chain, which results in high electrical conductivity, carrier transport has become simpler in the polymer chain. According to the IeV characteristics, the electrical conductivity of PANI/NiO NCs initially rises with the increase in NiO incorporation in PANI. Then, it begins to fall again after reaching a certain limit of NiO content. The electrical conductivity ($s$) with the greatest value was determined to be $1.39 \times 10^{-1}$ S/cm at 110°C for the PANI/NiO NC with 0.75% NiO addition. This value for electrical conductivity is around 17-times larger than that of pure PANI ($0.085 \times 10^{-1}$ S/cm).[13]

Similarly, the electrical conductivity of PANI can be enhanced by incorporating titanium dioxide ($TiO_2$) in the PANI matrix. The addition of the $TiO_2$ NPs in the PANI chain matrix created an ordered chain structure and a decrease in the carrier hooping barrier. This increased the charge carrier conduction, hence enhancing the TE properties of the material. The electrical conductivity of the PANI/$TiO_2$ composites increases from 0.000805 to 0.052 S/cm, with the weight percentage of $TiO_2$ varying from 0.5 to 1.5%. The formation of more conductive pathways is the reason behind the enhancement of conductivity.[14]

Recently, some research was conducted to check the possibility of extracting thermal energy from the human body, which can, in turn, improve the performance of wearable devices16. To demonstrate this, they created a flexible TEG (f-TEG) made up of $Bi_2Te_3$-based TE composite films on a flexible substrate. The output performance of approximately 0.38 μA and 2.3 mV was obtained from the biological output performance of the f-TEG on the human body.[15] TE paper created from bacterial cellulose/silver selenide (BC/$Ag_2Se$) NCs can be a promising candidate for the f-TEG17. The TE properties of these NCs show an enhancement with the electrical conductivity of 23,000 S/m and the Seebeck coefficient of −167 μV/K at 400 K.[16]

### 3.2.2 APPLICATIONS IN SOLAR PHOTOVOLTAIC CELLS

Solar photovoltaic cells work on the conversion of solar light energy to electricity via photovoltaic effects. The development of conventional silicon-based solar cells was constrained by their complex manufacturing process, high energy consumption, and high cost despite laboratory efficiency reaching more than 25%.[17] Thin-film solar cells, on the contrary, have disadvantages due to the scarcity of components such as Te, In, and Ga, as well as the toxicity of Cd and Se, which are frequently utilized in these cells.[18] Nowadays, extensive research is going on developing devices based on evolving photovoltaic techniques such as the following:

- Dye-sensitized solar cells (DSSCs),
- Perovskite solar cells (PSCs), and
- Organic photovoltaics (OPVs).

Polymer photovoltaic (PV) is another potential technology that uses polymer-based semiconducting organic molecules as solar absorber layers in solar cells. In comparison to other technologies, they have shown more reliability, such as lightweight, low cost, and flexibility.[19] A variety of nanomaterials encapsulated in polymer matrices have been investigated in the field of photovoltaics to address the rising demand for energy. A comprehensive discussion on various forms of PNCs reported for solar photovoltaic applications is given below.

#### 3.2.2.1 CARBON-BASED POLYMER NANOCOMPOSITES

NCs made of carbon allotropes and polymers have flexibility, low weight, and ease of manufacture that make them ideal for photovoltaic applications. They are extensively researched for use as electrodes, active layers, and electrolytes in various solar cells. In DSSCs, Pt is commonly used as a counter electrode owing to its good electric conductivity and excellent electrocatalytic activity. However, the applicability of Pt electrodes for large scale device fabrication is limited by its high cost, low abundance on earth, poor electrolyte medium stability, and high vacuum requirement for processing. Carbon allotropes and conducting polymer composites have recently recognized as an affordable substitute for Pt electrode. Carbon allotropes can act as conducting fillers for base polymer matrix to improve the catalytic performance of conductive/conjugated polymers.[20] As a replacement for Pt electrodes, PNCs made of MWCNTs and polythiophene were reported. These PNCs outperformed Pt-based solar cells in terms of photovoltaic performance. Similarly, polypyrrole (PPy) and functionalized MWCNTs composite film modified with Cu nanostructures was developed to replace Pt counter electrode in DSSCs.[21] The electrolyte in DSSCs is often an organic solvent that contains a redox mediator, such as an iodide/triiodide pair, for dye recovery and electrolyte regeneration during operation.[22] The primary drawback of liquid electrolytes related to their volatilization and leakage limits the usage of DSSCs in the commercial sector. To solve these issues, polymer quasi-solid electrolytes have been proposed as an alternative to liquid electrolytes. They have lower vapor pressure, greater adhesion and diffusion into the mesoporous anode, good thermal stability, and flammability.[23] They also have improved ionic conductivity due to the successful trapping of a significant volume of liquid electrolyte in the polymer matrix. They are, however, constrained by lower power conversion efficiency (PCE), which is attempted to be overcome by combining polymers with different plasticizers and forming composite with inorganic fillers. Sultana et al. have developed MWCNTs/ PANI NCs via an in situ polymerization process for use as electrolytes.[24] Using these PNCs as the host material, potassium iodide as the salt, and propylene carbonate as the solvent, a quasi-solid-state electrolyte was fabricated. A printable quasi-solid electrolyte is reported with $Ti_3C_2$ MXene-rGO composite in the copolymer polyethylene oxide/poly(vinylidene fluoride-co-hexafluoropropylene) (PVDF-HFP).[25] Furthermore, carbon-based PNCs are studied as the active layer in OPVs. Radaoui et al. have reported the use of a conjugated polymer poly(1,4-phenylene-ethynylene)-alt-poly(1,4-phenylene-vinylene)s in combination with fullerene as an active layer for solar cells.[26]

The main issues impeding PSCs' efficiency are related to lack of long-term device stability, perovskite defects, and loss of energy resulting from charge recombination. Carbon-based PNCs with multifunctional features can be used in PSCs in the form of the following:

a. charge transport layers,
b. electrodes,
c. perovskite additives, and
d. encapsulation layers to address the aforementioned problems.[27]

To address the low stability of PSCs, a protective coating with a multi-functional graphene/PANI composite was reported.[28] The $CH_3NH_3PbI_3$ perovskite crystals were encapsulated by this composite layer, which provided protection for perovskites vs degradation from oxygen, moisture, and also the UV light. Another function of carbon-based PNCs as perovskite additives is shown clearly in Figure 3.1.[29] In this figure, (PM6) composite or (PM7) composite, which is a very highly conductive graphene/Lewis base polymer, was used to passivate defects and to improve charge transfer in the perovskite film. A schematic representation of the device structure and its cross-sessional SEM image are given in Figure 3.1a, b, respectively. Now, based on the energy variation level of the device presented in Figure 3.1c, graphene/polymer composites provide an energy increment for hole transfer among perovskite and 2,2′,7,7′-tetrakis (*N,N-p*-dimethoxyphenylamino)-9,9′-spirobifluorene (spiro-OMeTAD), that in fact lowers the loss of energy and potentially raises the device's open circuit voltage. The molecular structures of polymers, PM6 and PM7, are represented in Figure 3.1d.

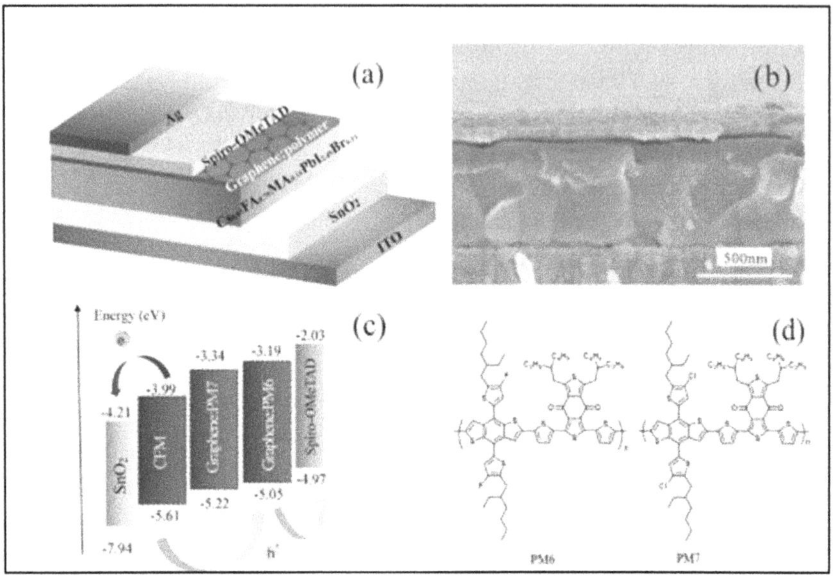


**FIGURE 3.1** (a) Assembly of the device, (b) SEM image showing the cross-sectional view of the device, (c) Energy level diagram of components of PSCs, and (d) Molecular structure of PM6 and PM7.

*Source:* Reproduced with permission from Ref. [29]. Copyright © 2021. American Chemical Society.

Zhou et al. reported the use of polyethylenimine (PEI) modified cross-stacked superaligned carbon nanotube (CSCNT) composite as electrodes in inverted PSCs.[30] The PEI (0.5 wt %) modification of CSCNT promoted interfacial charge transfer, enhancing the photovoltaic performance with a PCE of ~11%. In another work, an encapsulation layer made of poly(vinyl alcohol-co-ethylene) (EVOH) with $SiO_2$ and graphene oxide (GO) as fillers was reported.[31] Because of the improvement in the moisture barrier property, PSCs had great water stability and were able to uphold 86% of their performance even after being in direct contact with water for 5 h. The modification of the mesoporous $TiO_2$-based electron transport layer (ETL) with reduced graphene oxide (rGO)/PANI composite has also been reported.[32] This modification resulted in a PCE of around 16.48, 33.7% higher than that of pure mesoporous-$TiO_2$ devices. Yu et al. have demonstrated the use of GO-doped poly(3,4-ethlyenedioxythiophene): polystyrene sulfonate (PSS) composite film as a hole transport layer

(HTL) for promoting the stability and efficiency of inverted PSCs.[33] The composite film's superior optical and electrical characteristics, such as enhanced electrical conductivity, a higher work function, improved crystallinity of the perovskite crystal, and suppressed leakage current, resulted in excellent solar cell performance. Similar to this, rGO/poly(*N*-vinylpyrrolidone) NCs (rGO/PVP) also served as HTL in PSCs.[34]

### *3.2.2.2 METAL AND METAL OXIDE POLYMER NANOCOMPOSITES*

Many metal oxides have been utilized as inorganic ceramic fillers in the preparation of PNCs for photovoltaic applications. A quasi-solid electrolyte with high ionic conductivity and high photovoltaic performance was reported by Subramanian et al. for use in DSSCs.[35] Here, a polymer blend of PVDF-HFP and poly(methylmethacrylate) (PMMA) with 1D $TiO_2$ nanofiber filler was prepared by electrospinning method. The developed NC polymer membranes were activated using various redox agents such as LiI, NaI, KI, and tetrabutylammonium iodide (TBAI) to generate the quasi-solid electrolytes. $Fe_2O_3$ was also employed as filler in PVDF-HFP based polymer electrolyte, demonstrating PCE of 3.62% with a 2.1-fold improvement over pure PVDF-HFP electrolyte.[36] In another work, PVDF-based quasi-solid electrolyte was developed with rare earth metal oxide NPs and an ionic liquid.[37] For the fabrication of DSSCs, 3-decyl-1,2-dimethylimidazolium tetrafluoroborate ([dmim]$BF_4^-$) ionic liquid and dysprosium oxide particles were incorporated into PVDF by electrospinning. Here, the DSSCs with polymer-based electrolytes showed a higher conversion efficiency of 6.4% and better stability than liquid electrolytes. ZnO in a polymer blend matrix of PVDF-HFP and polyethylene oxide (PEO),[38] silica NPs surface engineered with PPy and PANI in polyurethane,[39] $Co_3O_4$ in poly(acrylonitrile)(PAN) host matrix blended with an ion-conductive poly(vinylpyrrolidone-co-vinylacetate)[40] etc. are some of the other examples for nanofillers used in combination with polymers in the preparation of polymer gel electrolytes.

In thin-film organic solar cells, a polymer blend comprised of poly(3-hexylthiophene) (P3HT) and [6,6]-phenyl C61 butyric acid methyl ester (PC61BM) was used as absorber medium.[41] To improve photon harvesting in a solar absorber medium, core-shell structured ZnS/Ag NPs were doped in it. In another study, to improve the efficiency of the solar cell, a photoactive layer composed of P3HT and [6,6]-phenyl-C61-butyric-acidmethyl-ester

(P3HT:PCBM) was modified with Ag/Zn NCs.[42] A novel superparamagnetic core-shell NC of poly(*m*-aminobenzenesulfonic acid) (PABS) and $Fe_3O_4$ denoted as NCPABS-$Fe_3O_4$ was developed for enhancing the performance and efficiency of polymer-based solar cells.[43] Here the PNCs were synthesized by *in situ* polymerization of *m*-ABS in the presence of $FeCl_3 \cdot 6H_2O$ as oxidant under solid-state conditions. The synthetic procedure and the incorporation of magnetic NC as a single layer in the fluorine-doped tin oxide (FTO)/$TiO_2$/NCPABS-$Fe_3O_4$/Al solar cell are schematically shown in Figure 3.2. In this case, $TiO_2$ deposited on FTO served as the anode and Al as the cathode. A thin layer of NCPABS-$Fe_3O_4$ composite was spin-coated onto the $TiO_2$ layer, and then the Al cathode was placed on top of the composite film. As shown in Figure 3.2, the complete assembly resembled a sandwich-like structure. The resulting solar cell showed a PCE of 4.24%, which was nearly 660% higher than those obtained from FTO/$TiO_2$/(PABS)/Al. Similarly, a superparamagnetic composite of PABS with two inorganic phases, $TiO_2$ and $Fe_3O_4$, prepared by solid-state polymerization, was used as an active layer in solar cells. The presence of superparamagnetic $Fe_3O_4$, acceptor $SO_3^-$ groups, and *N*-type $TiO_2$ NPs in the active layer resulted in a PCE of about 3 % under simulated solar irradiation.[44]

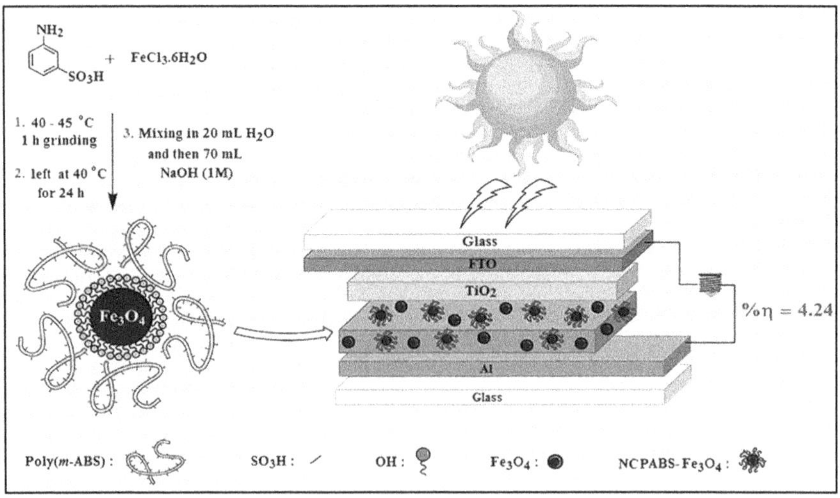


**FIGURE 3.2** Schematic representation of FTO/$TiO_2$/NCPABS-$Fe_3O_4$/Al solar cell assembly.
*Source:* Reproduced with permission from Ref. [43]. Copyright © 2020. Elsevier.

To fabricate an innovative polymer solar cell, an NC of poly(*m*-aminobenzodioxol) and $Fe_3O_4$ magnetic nanorods (PABD-$Fe_3O_4$NRs) was developed and used as the active layer.[45] For the encapsulation of the active layer, the composite was incorporated within a graphite-implanted polyacrylamide (PAA-G) hydrogel matrix. The extra $Fe^{3+}$ was utilized to create a metal thiocyanate complex, $[FeSCN]^{2+}$, inside the pores of the hydrogel. The entire system was designed to promote the development of interconnected channels, which increased the electrocatalytic area and reduced charge-transfer resistance. It is recognized that in HTL-free-PSCs, the device performance can be improved by adding suitable materials into the ETL. To enhance the perovskite crystalline structure, its performance, and address the instability problem of these cells, a bilayer ETL was created by modifying the mp-$TiO_2$/perovskite interface with copper oxide@polyaniline (CuO@PANI) *n*-composite.[46] This novel bilayer ETL increased the short-circuit current density and PCE while retarding the charge recombination.

### 3.2.2.3 *CONDUCTING POLYMER NANOCOMPOSITES*

Conducting PNCs have attracted more interest than traditional polymer-based materials because of their high electrical conductivity, facile synthesis, biocompatibility, large surface area, increased carrier transport, and distinctive optical properties.[47] Conducting polymers are employed as an electrocatalyst in solar cells and have high energy conversion efficiency but are less expensive than silicon-based solar cells.[48] As a result, conducting polymer-based solar cells are considered to be an excellent substitute for silicon solar cells. In the earlier sessions, some PNCs utilizing conducting polymers were discussed. PPy incorporated with $SrTiO_3$ nanocubes were used as counter electrodes in DSSCs.[49] The electrode developed with the addition of $SrTiO_3$ into the PPy matrix enhanced the electrocatalytic response to the redox system, improved the surface area, and exhibited low charge transfer resistance compared to a pure PPy electrode. Similarly, He et al. developed composite films of titanium carbide (TiC) and poly(3,4-ethylenedioxythiophene) (PEDOT) to use as counter electrodes in DSSCs.[50]

In polymer based OPVs, the deposition of thin films made of polymeric materials is very crucial during device fabrication. Many PANI derivatives, such as poly(ortho phenylenediamine) (PoPDA),[51] poly (ortho anthranilic acid),[52] poly (*ortho*-aminophenol),[53] etc., have been used for thin-film reparation. Recently, a terpolymer-blended thin film based on ortho phenylene diamine, meta phenylene diamine, and pyrrole terpolymers was reported to achieve enhanced optical conductivity for solar cells.[54] The thin film of terpolymer was prepared via the physical vapor deposition method and was characterized by a fibrous structure. In another study, the photovoltaic properties of thin films made of poly(3-octyl thiophene) (P3OT) doped with $Ni0.4Sr0.6TiO_3$ NCs were investigated as a function of P3OT in different concentrations.[55] They observed significant improvements in solar cell properties with 50% doping of P3OT, which was attributed to factors such as dopant concentration, type of solvent used, and morphology of the thin film.

### 3.2.3 POLYMER NANOCOMPOSITES FOR SUPERCAPACITORS AND BATTERIES

The demand for renewable energy resources is increasing day by day. Research that has been carried out in the past decade concentrates its findings on renewable energy resources and devices. The development of lithium-ion batteries (LIBs) is considered as one of the major milestones among them. Vast research was carried out to improve the quality and lifetime of these batteries. Nowadays, the limited availability of raw materials increases their cost. Their low lifetime and toxicity came as a challenge. There comes the development of alternatives such as Li- sulfur and Al-air batteries. Another important invention in the field of renewable energy storage is supercapacitors. They possess exceptionally good power density and can replace traditional batteries.[56] The charging/discharging process in supercapacitors is quite high when compared with the batteries in which polarization resistance is present.[57] Different types of materials can be used as electrode materials in supercapacitors and batteries. Carbon-based materials, metal, and metal oxide NPs are some of the examples for them. These materials possess high power densities and cycling performance. The limitations of such materials are their poor energy densities. Polymers such as PANI, PPy, etc. can be used as electrode materials in supercapacitors because of their high pseudo capacitance. But they possess poor cyclic stability. Figure 3.3 represents three types of supercapacitors and their building materials.

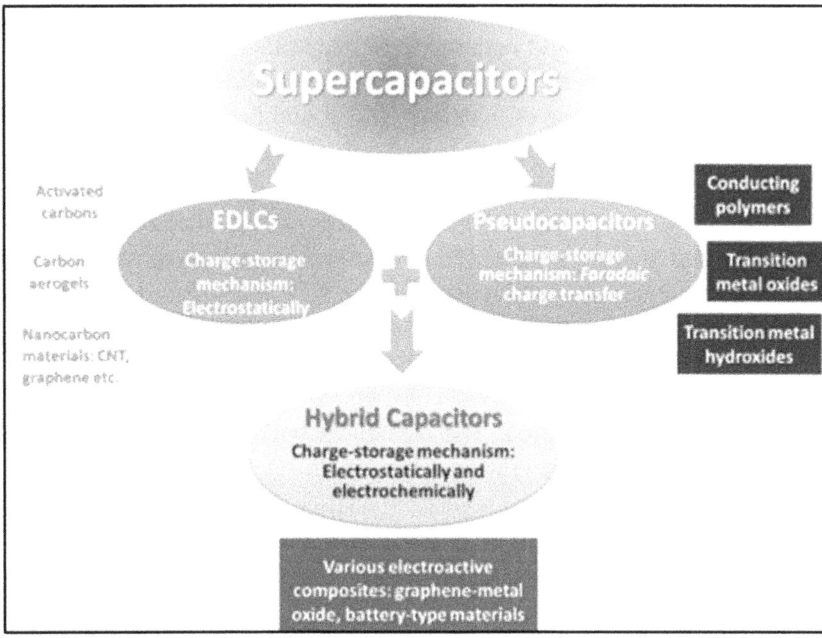


**FIGURE 3.3** Three types of supercapacitors and corresponding most common materials used in the device construction.

*Source:* Reproduced with permission from Ref. [57]. Copyright © 2017. Elsevier.

Thus, the incorporation of nanomaterials with polymers gives PNCs, and they can act as potential candidates for electrode materials in supercapacitors and batteries.[58] PNCs as electrode materials can improve transport parameters such as ion mobility and ion transport correlation in batteries.[59] PNCs can be prepared via different methods which can modify the applications of their parent materials. PNCs can be fabricated into the form of films, foams, and fibers with superior properties. PNCs of different nanofillers are discussed elaborately below.

### *3.2.3.1 CARBON-BASED POLYMER NANOCOMPOSITES IN BATTERIES*

Carbon-based PNCs are effectively used in supercapacitors and batteries as electrode materials. Activated carbon, graphene, graphene/fullerene binary hybrids, carbon-based ternary hybrids, CNTs, carbon dots, and carbon aerogels are used as electrode materials both in batteries and supercapacitors. The addition of carbon-based materials into a polymer matrix significantly increases surface-to-volume ratio and mechanical properties which includes flexibility of materials. It is found that the shape, size, and amount of loading of carbon fillers into polymers change the properties of NCs.[60]

PANI/CNTs composites can be prepared by chemical oxidative polymerization, and it shows tremendous electrochemical performance. It is seen that the coulombic efficiency of these NCs is increased to a greater extent when compared with the polymer alone. High coulombic efficiency is attributed to the improvement in reversible behavior of charging/discharging cycles. Similarly, it is also found that the discharge specific capacity also gets improved. These features will enable the material to act as good electrode material in LIBs. Figure 3.4 compares the curves of Coulombic efficiency and reversible charging/discharging behavior of PANI and PANI/MWCNTs composites. The improvement of features can be well identified from the graph itself.[61]

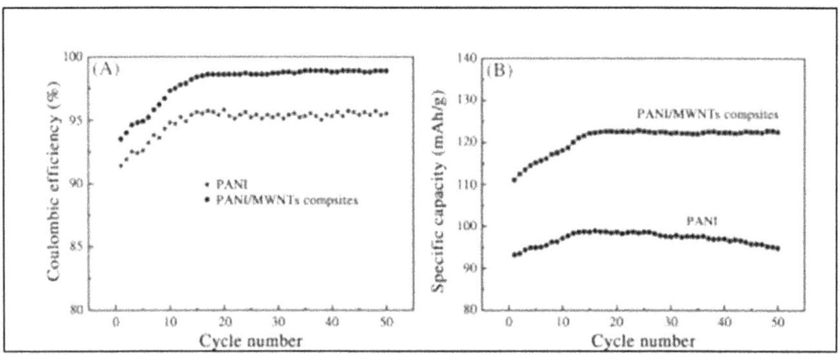


**FIGURE 3.4** (A) Coulombic efficiency and (B) cycle curves of the PANI and PANI/MWCNT composites.

*Source:* Reproduced with permission from Ref. [61]. Copyright © 2009. Elsevier.

Immense research is taking place in the field of supercapacitors and other energy storage devices. Carbon-based PNCs are effectively utilized in supercapacitors. Synthesis of PANI/graphene PNCs and utilization of this material as electrodes in supercapacitors are first done by Li et al. The synthesized materials possess high specific energy density and power density. Other PNCs of PANI and rGO exhibit a high specific capacitance of 782F/g and a current density of 1.14 A/g and are a perfect electrode material for supercapacitors.

PPy is another important polymer that is used to synthesize NCs with graphene and rGO for supercapacitor applications. Studies have also shown that doping in graphene or rGO with a nonmetal, such as nitrogen, will further increase the electronic transfer efficiency of composite materials. *N*-doped rGO/PPy electrode showed a specific capacitance of 394 F/g.[62] PANI-MWNT NCs prepared via the chemical oxidation polymerization method also act as an electrode material in supercapacitors since they possess a specific capacitance of 762 F/g.[63] A ternary composite of CNT/graphene/PPy also gives good performance as electrode material in supercapacitors.[64]

### *3.2.3.2 METAL AND METAL OXIDE POLYMER NANOCOMPOSITES IN BATTERIES AND SUPERCAPACITORS*

Transition metal oxide-PNCs act as good electrode materials in LIBs. Polymers such as polyacetylene, PANI, and PPy can be used to make composites with transition metal oxide like $V_2O_5$ to form hybrid materials with better electrochemical characteristics.[65] Oxides of iron, cobalt, tin, and titanium can be used to make PNCs which can act as anode materials in LIBs. A hybrid of graphene and $SnO_2$ is used to make NC with PANI shows exceptionally good electrochemical performance, and acts as an anode material in LIBs. Similarly, $TiO_2$-based NPs can also be employed as anode materials. However, the limitation is the poor conductivity of $TiO_2$ NPs, and it can be resolved by synthesizing a hybrid of $TiO_2$ nanohybrid with graphene nanosheets.[66]

PANI-based PNCs of $RuO_2$, NiO, $Fe_2O_3$, $NiCo_2O_4$, $MnO_x$, $MoS_2$, and $SnO_2$ exhibit high electrochemical performance and are thus utilized in supercapacitors as electrodes. Ag/Au nanowires with PPy are now widely used in stretchable supercapacitors. Similarly, silver nanowires with PEDOT polystyrenesulfonate (PEDOT:PSS) are utilized for developing a transparent, stretchable supercapacitor.[67] The development of PANI/$MnO_2$ PNCs enhances the effective surface area, thereby improving the charge/discharge rates in supercapacitors. It is found that the fibrous PANI/ $MnO_2$ gives a specific capacitance of 525 F/g. It possesses excellent cycling stability and current density. The material has superior properties, which is needed for an electrode material used in supercapacitors.[68] Another important metal oxide is NiO is used to make PNCs with PANI and shows a capacitance of 514 F/g at a scan rate of 1 mV $s^{-1}$.[69]

### 3.2.3.3 VERSATILE NATURE OF CONDUCTING POLYMER NANOCOMPOSITES IN BATTERIES AND SUPERCAPACITORS

Conducting polymers are widely used as pseudo-capacitor materials in flexible supercapacitors. They possess high redox active-specific capacitance and an elastic polymeric nature. Conducting PNCs offer next-generation energy system storage devices that are flexible, lightweight, and cost effective. Conducting PNCs are used in supercapacitors and rechargeable batteries.[70] PANI derived PNCs act as potential candidates in supercapacitors and batteries. The application of graphene-PEDOT NCs as an electrode material in supercapacitors attracts much attention in this field. G-PEDOT has both graphene and PEDOT characteristics and it can accommodate dopant ions in a better manner.[71] Polymers based on aromatic diamines provide much more novel properties than PANI and PPy. It is due to the free amino group per unit on polymer chains. One such example is poly(*p*-phenyldiamine) (PpPDA). PpPDA is an electroactive polymer whose structure is closely related to PANI. Studies have shown that PpPDA is widely used as electrode material in supercapacitors and batteries. The easy and unique method of synthesis of PpPDA also extends its applications.[72] The field of conducting PNCs for supercapacitor and battery applications is still in the area of vast research.

### 3.2.3.4 POLYMER NANOCOMPOSITES OF OTHER FILLERS

PNCs of fillers other than carbon-based material and metal NPs are also utilized in the field of supercapacitors and batteries as electrode material. One of the most promising candidates in this field is MXenes. MXenes are 2D materials that are transition metal carbides, nitrides, or carbonitrides, which are generally represented as $M_{n+1}X_n$, where M is a transition metal, and X is either oxygen or nitrogen. The incorporation of MXenes into polymers provides versatile materials with exceptionally good properties. The mechanical properties of MXene/PVA NC prepared by Ling et al. are very superior as compared with their parent compounds. Boota et al. synthesized PNCs of PPy and MXene, and the resulting compound possesses high volumetric capacitance density in the range of 1000 F $cm^{-3}$ and hence widely utilized in supercapacitors as electrodes. PANI, PEDOT, and polyfluorene derivatives (PFDs) also form NCs with MXenes, and the materials are actively evolved as electrode materials in supercapacitors and batteries. PNCs of PEDOT: PSS and MXene are used to prepare supercapacitors, and changing the reaction conditions may vary the capacitance of materials.[73] Synthesis of PPy-encapsulated polyoxomolybdate-decorated MXene helps for the development of electrodes in highly durable and great-performance Li-ion batteries. Figure 3.5 schematically represents the use of MXene/PPy PNCs in batteries and supercapacitors.[74]

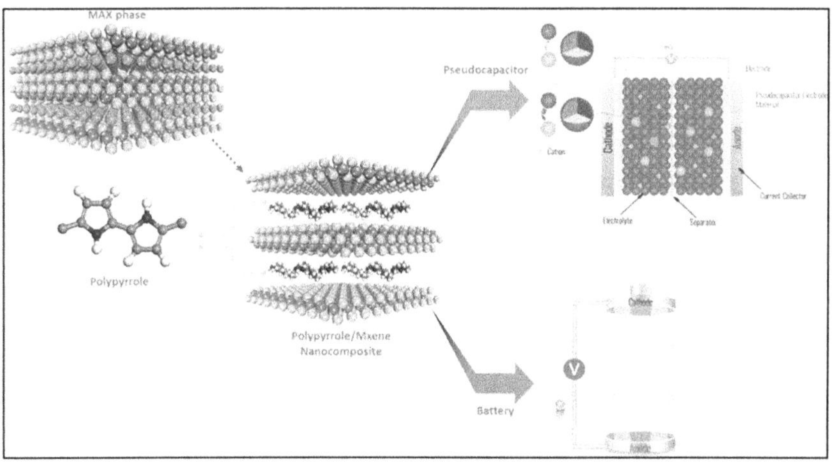


**FIGURE 3.5** Schematic representation of the formation of polypyrrole/MXene nanocomposite and use of the materials in supercapacitors and batteries.

*Source:* Reprinted from Ref. [74]. Copyright 2022. American Chemical Society.

### *3.2.4 FUEL CELL*

A fuel cell is, as a matter of fact, an electrochemical cell. This fuel cell converts the chemical energy of a fuel. This conversion is done with the use of an oxidizing agent into electricity through a pair of redox reactions.[75] Fuel cells have gained attention as potential replacements for a number of energy storage or power production systems due to their lesser environmental pollution than conventional power generation technologies.

### 3.2.4.1 CARBON-BASED POLYMER NANOCOMPOSITES FOR FUEL CELL

Activated carbon, graphene, CNTs, carbon dots, and carbon aerogels are examples of carbon nanomaterials that can be used as suitable fillers for polymeric materials and aid in improving the polymer's characteristics. Among the various types of fuel cells, the Proton exchange membrane fuel cells (PEMFCs) are suitable for small-scale power generation systems due to their low operating temperature and simple design.[76] PEMFCs and direct methanol fuel cells (DMFCs), which use polymer electrolyte membranes (PEMs) as electrolytes, are the most promising candidates for low-temperature operations, including mobile and portable applications.[77]

Carbonaceous nanomaterials like fullerene, CNTs, graphene, and carbon particles, as well as their chemically functionalized counterparts, continue to be sought-after fillers for the creation of a variety of NC proton exchange membranes (PEMs) with desirable properties for fuel cell applications like PEMFCs and DMFCs.[78] When operating PEMFCs above 100°C and with low RH%, the use of such nanoscale fillers in these membranes was primarily to improve the thermal stability and hydromechanical properties of PEMs, whereas the inclusion of such fillers improved the methanol barrier properties in DMFCs. These carbonaceous nanofillers' ability to disperse and retain water was greatly enhanced by functionalization, resulting in PEMs with high conductivity and low methanol permeability. Baker et al. [79] prepared a composite membrane consisting of a Nafion proton exchange ionomer and ceria-coated MWCNTs by a solution-casting method. The mechanical strength of the membrane is boosted by the presence of MWCNTs, and its chemical endurance is increased by the inclusion of ceria by scavenging free radicals. Without compromising electrical or ionic conductivity, composite membranes showed enhanced tensile strength and dimensional stability during hydration and maintained the same level of polarization performance in a fuel cell system. Its high durability also increases the economic viability of fuel cell technology.

Microbial fuel cells (MFCs) use living catalysts to create electricity from energy derived from commercial, sustainable, and domestic waste sources. The carbon-based PNCs can have a very significant role in the MFC's cathode.[80]

Yang et al.[81] created *N*-doped carbon aerogels (NCAs) using the controlled pyrolysis of polyacrylonitrile at temperatures between 600 and 900°C. The resultant NC was applied in the MFC for the oxygen reduction process (ORR). The MFC produces electricity as well as is used to treat wastewater. According to electrochemical experiments, the NCAs appear to have electrocatalytic activity toward ORR in alkaline conditions. With a half-wave potential of +0.80 V vs RHE and a number of electron transfers of about 3.9, the sample synthesized at 800°C performed the best among the series. According to Liang et al., the MFC that had CNT particles added to it maintained an anodic resistance of around 180 Ω. The anodic resistance of those lacking CNT powders, however, rose by up to 540 Ω. This proved that the addition of CNT powders helps stabilize the anodic resistance and hence stabilize the power generation and internal resistance during long-term operation. Additionally, it was found that an MFC with spin-spray layer-by-layer CNTs that were horizontally aligned had thicker biofilm and lower sheet resistance in comparison to MFCs with CNTs that were vertically and randomly oriented.[82]

### *3.2.4.2 METAL AND METAL OXIDE POLYMER NANOCOMPOSITES FOR FUEL CELL*

Metal oxides are a viable solution to the current issues in polymer electrolyte fuel cells due to their accessibility, low cost, excellent chemical and electrochemical stability, abundance of hydroxyl groups on the surface, and strong interaction with metal NPs.

Lin et al.[83] synthesized the magnetite NPs on GO ($Fe_3O_4$@GO). In this present work, a polymer matrix is incorporated into it to produce an effective electrolyte membrane for direct alkaline fuel cells. In the film drying step, $Fe_3O_4$@GO nanofillers are distributed along the direction in a quaternized polyvinyl alcohol (QPVA) membrane by applying an external magnetic field. It results in a polymer-rich region in the center and a nanofiller-rich region near the membrane surface. The NC membrane reduces the permeability as well as improves the ionic conductivity. When the fuel cell is supplied with 2 M methanol or 3 M ethanol at 60°C and utilizing the magnetic-field assisted NC electrolyte, maximum power densities of 200 and 144.3 mW $cm^{-2}$ are reached. The fabrication of NC membrane electrolytes using a magnetic field has a significant role in alkaline fuel cells and energy-based applications.

A NC fabricated from NiO, PPy, and PANI with manganese oxide shows higher power production and waste treatment efficiencies.[84] It is because of their unique rich structural variety, lack of toxicity, environmental friendliness, high chemical activity, biocompatibility, and interaction specificity.

The properties of manganese oxide such as high thermochemical stability, elasticity, biocompatibility, better tensile strength, large surface-to-volume ratio, high loading capability, excellent penetrability, and promoting excellent electron-transfer reactions make it a promising biocatalyst functional material for energy harvest. So, the development of MFCs like such systems results in power efficiency enhancement with increasing electron transfer rate in any bioelectrochemical reactions. The plain graphite (PG) electrode coated with $MnO_2$/MWCNT acts as an anode in MFCs. The addition of 50% of $MnO_2$ improves the wettability of the electrode and decreases the contact angle. This will result in improved kinetic activity and power density of MFC, compared to unmodified ones.[85] The catalytic efficiency of perovskite oxide NC based $LaCoO_3$ (LC), $LaMnO_3$ (LM) and $LaCo_{0.5}Mn_{0.5}O_3$ (LMC) as cathode catalysts are tested in a dual-chambered MFC. In the two-chambered MFC, $LaMnO_3$ exhibits a maximum power density of more than 13.9 $mWm^{-2}$, which is twice as high as the cathode's highest power density when using carbon cloth (CC).[86]

#### 3.2.4.3 CONDUCTING POLYMER NANOCOMPOSITES FOR FUEL CELL

Even though conducting PNCs have only been around for ten years, they are highly important additions to the list of innovative materials and composites. Conducting polymers have been employed as the support matrix in various composites for the intercalation of catalytically significant NPs to preserve the composite's catalytic activity.

Jaafar et al.[87] incorporated conductive PANI-deposited oxidized graphitic carbon nitride (PANI-O-g$C_3N_4$) in a sulfonated poly ether ether ketone (SPEEK) membrane for DMFC applications. The morphological and X-ray diffraction studies show that the structure and crystallinity were changed with the deposition of PANI on g-$C_3N_4$. The modified SPEEK membrane improved not only the water uptake capacity but also the ion exchange capacity. Hence, it is a superior nanofiller for the modification of PEM. In the creation of low-cost PEM for portable DMFC application, conductive polymer-coated O-g-$C_3N_4$ at higher loading concentrations may be employed as an efficient methanol barrier filler. The electrochemical properties can be enhanced by incorporating the different amounts of sulfonated graphene oxide (SGO) into a PEM consisting of sulfonated poly ether sulfone (SPES). The incorporation of SGO enhanced the properties of the proton exchange SPES membrane, such as ion exchange capacity, proton conductivity, and methanol permeation resistance; SGO/SPES-based highly conducting polymer electrolyte membranes are used in the application of fuel cells.[88] Muthumeenal et al.[89] incorporated GO into the SPES membrane, which improved the proton conductivity and the mechanical properties and blocking of methanol molecules passing through the membrane. The performance of SPES/GO membranes was 3-times better than that of the original SPES membrane, making them a strong option for application in DMFC to address the present methanol crossover problems. Elakkiya et al.[90] modified SPES with a $TiO_2$ PEM. The surface-modified $TiO_2$- PANI NC shows the highest proton conductivity value of $2.30 \times 10^{-4}$ S/cm. Also, the sulfonic acid and amine group enhanced the ion exchange capacity and exhibited better fuel cell properties than the pristine SPES membrane. For the creation of a composite membrane in the fuel cell application, it would be a promising conductive material.

Zhang et al.[91] demonstrated the synthesis of AuNPs/PEDOT:PSS NC. The characterization studies show that AuNPs sizes within the PEDOT:PSS matrix depend on the plasma treatment time and precursor concentration, affecting the NCs' electrical conductivity and their catalytic performance. As they are dispersed in a continuous polymer film, they enhance the electron transfer efficiency. The studies show that it is a promising candidate for fuel cell catalysts.

#### *3.2.4.4 POLYMER NANOCOMPOSITES OF OTHER FILLERS*

The PNCs incorporated with MXenes are promising candidates for fuel cell applications. Elancheziyan et al.[92] developed an electrocatalyst using PANI/palladium/$Ti_3C_2T_x$ (PANI/Pd/MXene) NC for the methanol electrooxidation reaction. The chronoamperometric cyclic stability analysis shows that $Ti_3C_2T_x$-PANI-PdNP/SPE shows no reduction in the presence of methanol compared to $Ti_3C_2T_x$-PdNP/SPE. Figure 3.6 shows the cyclic stability of $Ti_3C_2T_x$-PdNP/SPE NCs.

The electroanalytical studies show the enhanced electrocatalytic response toward the oxidation of methanol. It is due to the metal–support interactions between PANI/Pd and MXene nanosheets and the efficient charge transfer from the MXene surface to the PANI/Pd electrocatalyst. So, MXenes are suitable candidates for methanol electrooxidation reactions in direct methanol fuel cells.

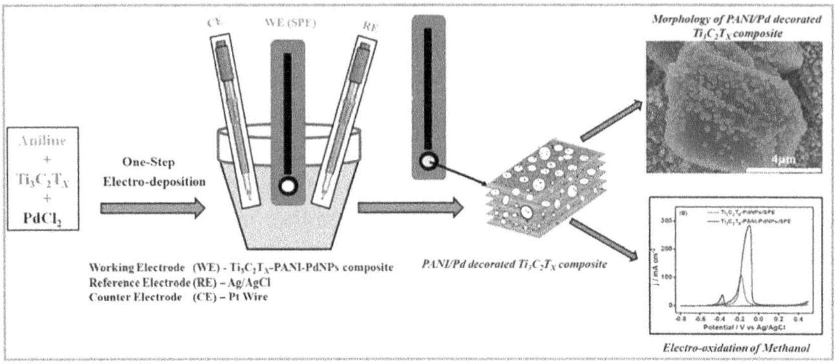


**FIGURE 3.6** $Ti_3C_2T_x$-PANI-PdNP/SPE nanocomposite methanol electrooxidation reaction.

*Source:* Reproduced with permission from Ref. [92]. Copyright © 2021. Elsevier.

## 3.2.5 HYDROGEN STORAGE

Hydrogen is a clean and renewable source of energy. Storage of hydrogen through gas compression, hydrogen liquefaction, or in the form of metal hydrides has significant drawbacks, including being too costly, too heavy, taking up too much space, or dangerous. Since nanotubes are chemically stable and have a low mass density, these disadvantages have prompted the study of new storage methods such as physical adsorption on nanomaterials such as graphite nanofibers, single-walled nanotubes (SWNTs), and multiwalled nanotubes (MWNTs).[93] Further nanomaterials blended with polymers to form composites like carbon-based PNCs, metal/metal oxide-based PNCs, conducting PNCs etc. can be used for effective hydrogen storage applications.

### 3.2.5.1 CARBON-BASED POLYMER NANOCOMPOSITES FOR HYDROGEN STORAGE

In the synthesis of a water-soluble polymer–SWCNT based composite for hydrogen storage, water-soluble polymers such as polyvinyl alcohol (PVA) and polyvinyl pyrrolidone (PVP) are chosen as the base material for hydrogen storage. Thin films made from composite solutions using the spin coating technique and the acid-treated polymer films are then hydrogenated at 50°C. For the acid-treated PVA and PVP, respectively, the

$Cl^-$ ions in the polymer matrix resulted in storage capacities of 0.9 and 1.3 wt %, whereas the storage capacity of composite films like PVA-SWCNTs and PVP-SWCNTs is 1.2 and 1.5 wt %, respectively. The desorption temperature ranges of the hydrogenated composite films PVA-SWCNTs and PVP-SWCNTs are determined to be 175–195°C and 120–145°C, respectively.[94]

Another research on polyetherimide (PEI)-based NC materials for hydrogen storage was reported. Here, MWCNT doped with sodium alanate ($NaAlH_4$) was produced as NPs, and PEI was used as the polymer matrix. The CNTs in the polymer matrix are poorly dispersed and have a highly aggregated morphology. At 120°C, $TiO_2$-doped PEIS/$NaAlH_4$ can store 1.1 wt % hydrogen. Although the increase in porosity of the material generates a little improvement in its hydrogen sorption capabilities, the addition of CNTs to the composite PEIS/$NaAlH_4$ has little impact on the material's hydrogen sorption properties.[95]

PANI-$H_3BO_3$/carbon nanotubes (PANI-B/CNT) composite and PANI-$BF_3$/carbon nanotubes (PANI-$BF_3$/CNT) composite are potential candidates for hydrogen storage. It was reported that the hydrogen adsorption capabilities of these materials were modified within the range of 0.44–3.67 wt % at the liquid nitrogen temperature and gas pressure up to 100 bar. Increasing the pressures up to 100 bar induced an increase in the hydrogen capacities of the materials within the range of 2.15–9.57-fold.

Hydrogen adsorption is also significantly influenced by the kind of carbon nanotube and the presence of organic functional groups during composite manufacture. At 100 bar and -196°C, the amounts of hydrogen in weight percent for SWNT (single-walled carbon nanotube), PANI-B, and PANI-$BF_3$/SWNT composites were respectively 3.01, 3.36, and 3.67, whereas these values for MWCNT, PANI-B, and PANI-$BF_3$/MWNT composites were 1.27, 2.24, and 2.73. In terms of hydrogen absorption per unit mass, SWNTs outperform MWNTs. At low temperature (−196°C) and high pressure (100 bar), the PANI-BF3/SWNT composite achieved the maximum H adsorption capacity of 3.67 wt %.

#### *3.2.5.2 METAL/METAL OXIDE POLYMER NANOCOMPOSITES FOR HYDROGEN STORAGE*

An air-stable composite material made of metallic Mg nanocrystals (NCs) in a matrix of gas-barrier polymer that allows for both high-density hydrogen storage (up to 6 wt % of Mg, 4 wt % for the composite) and loading in around 30 min at 200°C was investigated by Jeon et al.[96] Additionally, the Mg's nanostructuring enables quick storage kinetics without the use of costly heavy-metal catalysts. PMMA polymer has selective gas permeability, protecting the NCs from $O_2$ and $H_2O$. New possibilities in low-cost, high-capacity hydrogen storage media, batteries, and fuel cells are opened up by this method of creating nanosized, air-sensitive metal nanocrystals that are enclosed in a gas-selective polymer.

Another polymer-stable magnesium NC can also be prepared by laser ablation for efficient hydrogen storage. The Mg-NPs are enclosed in a polymer matrix of PMMA, which allows them to take up hydrogen more quickly than other Mg-pure kinds (around 20 min at 250°C) and with a higher capacity (6 wt % in Mg, 5.5 wt % overall). Figure 3.7 represents the schematic diagram of a hydrogen storage device using nano-Mg/PMMA composites.[97]

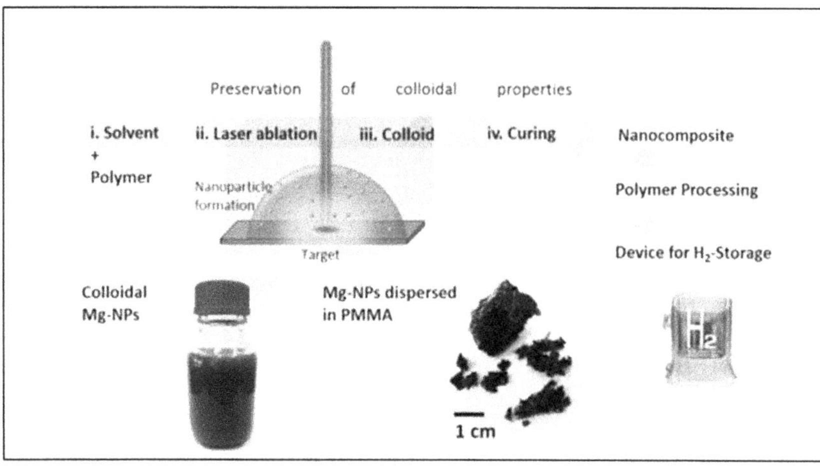


**FIGURE 3.7** Hydrogen storage using polymer stable magnesium nanocomposites prepared by laser ablation.

*Source:* Reproduced with permission from Ref. [97]. Copyright © 2013. Elsevier.

A series of NCs with varied percentages of Mg and polymer identities were tested to determine the impact of encapsulating polymer on nanocrystalline- Mg air stability and hydrogen storage density. In this study, a synthetic method to produce an Mg composite material that improves the hybridization kinetics by nanostructuring in which 5 nm spherical Mg nanocrystals are enclosed in a polymer matrix PMMA (120K MW) as a gas-selective barrier, which permits $H_2$ to enter while obstructing water and oxygen from reaching the NCs.

It was found that the overall properties of NCs do not scale directly with the individual component abundance (measured in terms of volume fraction of inclusion). Instead, the effects of nanoscale interactions on the functionality of macroscale materials are both unexpected and advantageous. An advantage of NCs that enhance polymer-nanocrystal interactions is that they increase both theoretical and experimental hydrogen sorption capacity as well as total air stability against oxidation. To achieve this, less amount of PMMA polymer is used, and more active material (Mg) is used. Concerns about polymer breakdown and overall composite stability may arise from the high temperatures required for hydrogen absorption/desorption cycling.[98]

Further, sulfonated poly (etherimide)/aluminum NCs were developed for hydrogen storage. The sulfonated polymer structure was improved by the addition of zeolite and aluminum during the synthesis of the NC but did not alter its thermal characteristics, such as the glass transition temperature or the temperature at which degradation begins. By using TEM analysis, it was found that the zeolite and aluminum particles are present but poorly scattered and distributed on the surface and inside of the sulfonated poly (ether imide). In tests for hydrogen sorption, the amount of hydrogen that the composite PEIS/ZEO/Al adsorbed compared to the composite without aluminum was much higher. The polymer matrix and nanofillers used in this study have a great deal of promise to create novel hybrid materials with increased hydrogen storage capacity.[99] Another composite of PEI-$LaNi_5$ films showed 0.6% hydrogen sorption without employing any special activation procedures.[100]

Kim et al. studied the enhancement of hydrogen storage capacity in PANI-vanadium pentoxide NCs. At 70 bar/77 K, volumetric hydrogen storage measurements were made on pure VO (vanadium pentoxide), untreated PANI, and PANI-vanadium pentoxide NC (PANI-VONC) nanosheets. At liquid nitrogen temperature, high-pressure hydrogen sorption isotherms of PANI-VONC were performed. PANI-VONC (~1.8 wt %) has a significant increase in hydrogen storage capacity compared to pristine vanadium pentoxide (~0.2 wt %) and untreated PANI (~0.2 wt %). It is suggested that the intercalation of PANI into the layers of vanadium pentoxide is the process that causes the improvement in hydrogen storage in PANI-VONC.[101]

### 3.2.5.3 CONDUCTING POLYMER NANOCOMPOSITES FOR HYDROGEN STORAGE

The conductivity of PANI, which is in the order of 10 S/cm, makes it a conductive polymer. This is greater than the average nonconducting polymer value but far lower than the value of metals.[102] Aside from being conductive, PANI is also affordable and simple to polymerize. Therefore, it was chosen as a matrix material for the NC structure due to its simplicity.

Hybrid porous materials with $NaAlH_4$ are a potential hydrogen storage hydride, and PANI or sulfonated PEI can act as the polymer matrix. $TiO_2$ and multiwall carbon nanotubes were also added to the $NaAlH_4$ to increase its ability to absorb hydrogen. The thermal parameters of the polymer, such as the glass transition temperature and the degradation temperature, were modified by the NPs, which also had an impact on the polymer's structural makeup. A significant amount of hydrogen (up to 1.2 wt %) was found in the NCs after 12 h at a relatively low temperature (120°C) and 32 bar, according to the kinetics of hydrogen sorption tests. However, PANI/$TiO_2$/$NaAlH_4$ NCs, after 6 h at 120°C and 32 bar, produced a hydrogen sorption capability of 0.5 wt % of $H_2$. It showed an improvement of 67% over neat PANI, which has 0.95 wt % of $H_2$ after 12 h.[103]

In another study, electrospun PANI nanofibers were used for reversible hydrogen storage. PANI has been researched for hydrogen storage due to its numerous physisorption and chemisorption sites. A hydrogen capacity of 3 wt % was attained at a temperature of 50°C, whereas different hydrogenation cycles resulted in a capacity increase of at least 2-fold (6-8 wt %) at a temperature of 100–125°C. Further, in electrospun PANI fibers, reversible hydrogen storage capacities of 3–10 wt % were also discovered throughout a range of temperatures (25–125°C). Significant changes in the microstructure (nanofibrillar swelling effect) can be seen in the surface morphologies of these PANI fibers before and after hydrogen sorption, which is likely to increase the volumetric hydrogen storage capacity.[104]

Later a facile method was developed to produce PANI-nanofiber-polypyrrole (PANI-NF-PPy) composites with network morphology. The created PPy layer's thickness was adjusted using various oxidant concentrations and the hydrogen storage capacity was examined at room temperature, which revealed a two-fold increase in the novel composite materials' hydrogen uptake. The hydrogen gas storage capacity of the PANI-PPy composites is 0.91 wt %, compared to 0.46 wt % for the HCl-doped PANI nanofibers.[105]

Inorganic nanotube composites based on PANI were utilized as potential room-temperature hydrogen storage material. In comparison to the halloysite nanotubes' (HNT) capacity of less than 0.019 and the 0.05 wt % of PANI, the NC exhibits an increased hydrogen storage capacity of 0.78 wt % with outstanding reversibility. Numerous chemisorption sites for the absorption of hydrogen were formed during the solid-state milling of PANI with the HNTs. The hydrogen storage capacity of the NCs, which ranged from 0.22 to 0.78 wt % and peaked at room temperature and a pressure of 0.5 MPa, was controlled by the reduced lumen diameters.[106]

#### *3.2.5.4 OTHER FILLERS*

Nareshmuthu et al. reported the hydrogen storage capability of SPEEK-hexagonal boron nitride (h-BN) (SPEEK-h-BN) NC membranes. Phase inversion approach is used to construct the NC membranes while considering different amounts of h-BN (0, 1, 3, and 5 wt %). The poly-ether-ether-ketone has a 65% sulfonation level, according to proton nuclear magnetic resonance (1H NMR) spectroscopy. The SPEEK-5% h-BN membrane performs better than pure SPEEK membrane, with a maximum storage capacity of 2.98 wt % at 150°C and an average binding energy of 0.38 eV for the adsorbed hydrogen. Due to the greater number of adsorption sites that h-BN offers, the SPEEK-h-BN NC membrane's capacity to store hydrogen rises with increasing h-BN. At room temperature, the hydrogenated membranes are stable. The stored hydrogen was found to release at a temperature of 214–218°C, and the SPEEK-h-BN membrane exhibits 100% desorption at this temperature.[107]

### ***3.2.6 ENERGY APPLICATIONS OF PIEZOELECTRIC POLYMER NANOCOMPOSITES***

Piezoelectric polymer composites are a class of materials having a filler with high piezoelectricity impregnated in a polymer matrix, which helps to combine the advantages of both materials; flexibility of polymers, elevated piezoelectric constant and coupling factor of fillers. The incorporation of fillers affects the all-intrinsic properties of the resulting piezoelectric NCs. The effect of filler polymer interfaces in PNCs has been elaborated by Prateek et al.[108] The amount, type, shape, and morphology of various fillers used in the construction play a considerable part in increasing the piezoelectric properties of NCs. Generally, fillers are classified into three major categories, namely, conducting, nonconducting, and hybrid fillers. Nanostructured piezoelectric materials, such as polymers, PNCs, and piezoelectric films, are utilized for many energy-related applications, especially for energy harvesting, energy storage, and energy conversions.

### *3.2.6.1 ENERGY HARVESTING APPLICATIONS OF PIEZOELECTRIC POLYMER NANOCOMPOSITES*

The size, structure, and comparatively long-term stability of the piezoelectric nanogenerators (PENGs) enable them to harvest energy by utilizing energy sources from our living surroundings, even from the mechanical impacts. Among the various piezoelectric materials, lead zirconate titanate (PZT), lead magnesium niobate-lead titanate (PMN-PT), barium titanate ($BaTiO_3$), etc., exhibit good level of piezoelectric response. PENG are made by incorporating ceramic nanofillers of high piezoelectric coefficients in various flexible polymeric media of good processability.[109] Piezoelectric energy harvesters, which convert an input of mechanical energy into electrical energy output, have been of great interest during the recent decade for self-powered and ubiquitous applications.[110] Zhang et al. prepared a rotational piezoelectric energy harvester that utilized wind energy by induced vibration of a PVDF beam.[111]

#### ***3.2.6.1.1 Nanocomposites Based on Nonconducting Fillers***

Among various polymer composites, polyvinylidene fluoride (PVDF)-based and NCs- based composites have been considered to be the superior option because of the greater piezoelectric coefficients of PVDF when compared to other polymers.[112] Bhunia et al. fabricated flexible ZnO/PVDF composite films by the sol–gel technique, followed by poling them in a vacuum.[113] Upon poling the ZnO/PVDF NCs, a 10-times improvement in output voltage was observed. Shin et al. prepared a NC film of hemi spherically aggregated $BaTiO_3$ NPs and P(VDF-HFP) to construct a PENG.[114] The construction of integrated $BaTiO_3$-P(VDF-HFP) clusters increased the piezoelectric output of the PENG by amplifying the total dipole moment and fostering a suitable structure for harvesting external stress. Yaqoob et al. assembled a PENG using PVDF/$BaTiO_3$-55 wt % NC and surface operationalized *n*-type graphene (nGr).[115]

Shetty et al. fabricated a PENG using electrospun-talc/PVDF nanofibers and proved that 0.5 wt % talcs caused the nucleation of 89.6% β-phase.[116] Park et al. prepared an NC by impregnating $BaTiO_3$ NWs into the polydimethylsiloxane (PDMS) matrix without using any dispersion enhancers.[117] Jung et al. constructed a flexible PENG using $NaNbO_3$ NWs/PDMS composite in a volume ratio of 1:100 between Au/Cr-coated Kapton films, and this kind of PENG exhibited an output voltage of 3.2 V due to the large $d_{33}$ coefficient of perovskite $NaNbO_3$ NWs.[118]

#### 3.2.6.1.2 Nanocomposites Based on Carbon-Based Fillers

There are several piezoelectric PNCs that serve profound applications in the field of energy that use a carbon-based conducting filler associated with it, namely, Ag[119]MWCNTs[120] rGO[121] and carbon black.[122] A voltage of 2 V was obtained from the nanofibers with 0.4 wt % Ag due to the well-aligned β-phase present in the samples as a result of the electrospinning process. The electrospun NCs showed a high β-phase amount as a result of straining the PVDF jets immoderately. The Ag-CNTs/PVDF fibers showed the largest $d_{33}$ value of 54 pC N–1 compared to 35 pC N–1 for PVDF/CNT Fibers and 30 pC N–1 for clean PVDF.

Alamusi et al. fabricated rGO/PVDF and rGO/PVDF films containing 0.05 wt % rGO displayed a peak output voltage of 3.28 V, around 293% of pure PVDF film. The main reason behind this enhanced output was the formation of a higher β-phase amount at 0.05 wt % rGO concentration, as confirmed by X-ray diffraction (XRD) and Fourier-transform infrared spectroscopy (FTIR) analyses.[121] Karan et al. reported a PEH using nonpoled iron-doped rGO/PVDF NCs with 2 wt % Fe-rGO loading generated a maximum $V_{OC}$ of 5.1 V as a result of continuous imparting by fingers. This enhanced result was achieved because of delocalized π electrons, electrostatic interactions between $-CH_2-$/$-CF_2-$ dipoles in PVDF, and oxygen functional groups in Fe-doped rGO through ion–dipole and hydrogen bonding interactions.[123]

### *3.2.6.1.3 Nanocomposites Based on Conducting and Nonconducting Filler Combination*

There are classes of NCs synthesized using a combination of conducting and nonconducting fillers and hybrid fillers such as $NaNbO_3$-rGO[124] manganese oxide ($MnO_2$)/graphene/MWCNT hybrids[125] titanium oxide ($TiO_2$)-rGO nanotubes.[126] Ponnamma et al. introduced a combination of 1D $TiO_2$ nanotubes, 2D rGO, and 3D strontium titanate into the PVDF-matrix followed by poling. The output voltage was about 10.5 times greater than the voltage produced by a neat PVDF matrix.[126,] Abbasipour et al. investigated the effect of three nanofillers, namely, graphene, GO, and halloysite nanotubes with varying concentrations, on piezoelectric properties of PVDF. The association of these nanofillers enhanced it further by 10%. Owing to the weak interactions of graphene with PVDF, it had a low impact on the piezoelectric properties of the NCs.[127] Highly aligned and thinner nanofibers were obtained from halloysite nanotubes because of their rod-like morphology when compared to PVDF/GO and PVDF/graphene nanofibers. A PENG was manufactured by Park et al. using a NC of $BaTiO_3$ NPs; SWCNTs or MWCNTs; or rGO dispersed into PDMS and generated an output voltage of ≈3.2 V.[128] Sun et al. developed a flexible PENG consisting of ZnO NPs/MWCNTs/PDMS NCs and the resulted in high output voltage. The association of MWCNTs significantly enhanced the PENG's voltage output due to the increased conductivity of the NCs. Park et al. added PZT particles and MWCNTs into PDMS and showed that PZT particles were well entangled with MWCNTs; the PZT particles showcased excellent crystallinity and exhibited outstanding piezoelectric properties.[129]

### *3.2.6.2 ENERGY STORAGE APPLICATIONS OF PIEZOELECTRIC POLYMER NANOCOMPOSITES*

Different energy storage devices have their individual characteristic applications. The characterization of NC for their structural and morphological studies is observed by FTIR studies and SEM to confirm the evolution of physiomechanical, thermal, and barrier properties of polymeric NC for energy storage application. PVDF-ceramic composite systems that combine the flexibility, elasticity, and stiffness of ceramics are expected to exhibit improved piezoelectric performance than that of neat PVDF. Ceramic fillers with high dielectric permittivity are a general choice to improve the dielectric and energy storage performance of PVDF-based composite systems. The addition of PLZT to PVDF and further CNT in PLZT-PVDF composite resulted in the enrichment of piezoelectric and energy storage properties. the amount of power delivered by the fabricated nanogenerator can be used for powering small electronic devices.[130]

PVDF as a dielectric polymer matrix combined with graphene platelets as a conductive and barium titanite as ceramic ferroelectric fillers were used to analyze the enhancement of energy storage ability at different graphene/barium titanate/ PVDF NC ratios by changing the concentration of graphene and barium titanate. The results showed that a combination of graphene and ferroelectric ceramic additives is an excellent approach to advance the performance of dielectric and ferroelectric properties of piezoelectric PNCs for energy storage applications.[131] $TiO_2$ NPs embedded Î$^{3}$-phase containing PVDF NC film turns to an excellent material for energy storage application due to an increased dielectric constant, and owing to their electroactive and extraordinary mechanical properties, PNC films have a strong ability to fabricate the PENGs.[132]

### *3.2.6.3 ENERGY CONVERSION APPLICATIONS PIEZOELECTRIC POLYMER NANOCOMPOSITES*

Piezoelectric systems are well appropriate to microsystem manufacturing technologies (MEMS).[133] A piezoelectric NC is designed to transduce daily human activities and vibrations into usable electrical energy. When mechanical energy is applied to the piezoelectric polymer film, electric charges are induced. With this property, a piezoelectric material can be considered as an electromechanical energy transducer.[134] The hybrid PENG shows high output power density and energy conversion efficiency and powers up many portable electronic devices like wristwatches, calculators, and mobile liquid–crystal display (LCD) screens through capacitor charging.[135] Lead-free alternative materials for converting mechanical energy into electrical energy through piezoelectric transduction are of significant value in a diverse range of technological applications.[136]

Nowadays, conversion efficiency of piezoelectric NCs with carbon-fiber-reinforced polymer electrodes are effectively utilized for stress sensing.[137] The PVDF-tungsten disulfide [$WS_2$] composed flexible device is able to produce a high output voltage of ~116 V and a piezoelectric energy conversion efficiency of ~25.6%, which is the highest among the reported values for PVDF-2D material-based PENGs. The uniformity in distribution of $WS_2$ nanosheets in the PVDF matrix accelerated the nucleation of the piezoelectric polar β-phase in PVDF-$WS_2$ NC and the piezoelectric performance of the PVDF-$WS_2$ based PENG is enhanced and is found to be superior to that of the other such systems.[138] Schematic mechanisms of (a) interactions between WS2 and PVDF and (b) operation of a piezoelectric nanogenerator and the device image are shown in Figure 3.8. A special type of nanogenerator generated a voltage output of ~30 V and current output of 3.7 μA with high stability and reproducibility using electrospun PVDF/PANI/graphitic carbon nitride nanosheets (g-$C_3N_4$) blend NC fibers (PPBF). The PPBF nanogenerator, which was ultraflexible and durable, exhibited high power density and conversion efficiency and was able to light up approximately 70 commercial light-emitting diodes (LEDs).

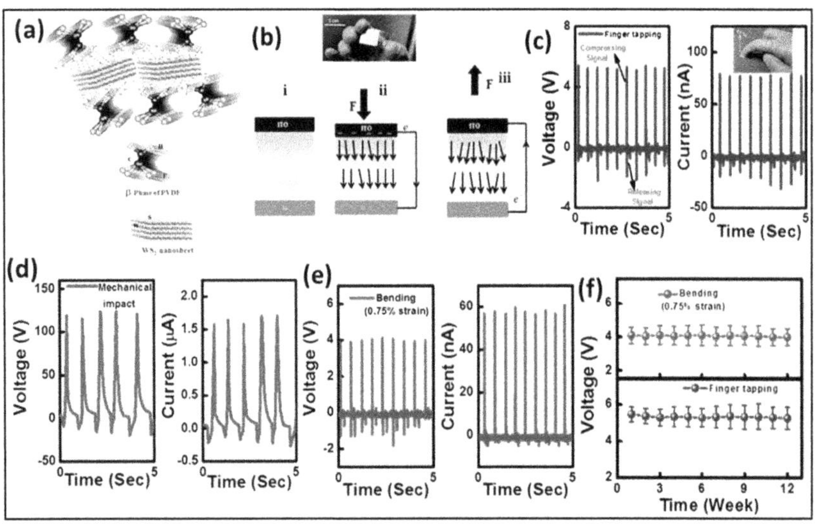


**FIGURE 3.8** Schematic mechanisms of (a) interactions between WS2 and PVDF and (b) operation of a piezoelectric nanogenerator, and the device image is shown in the inset. Open-circuit output voltage and current from the PVDF-WS2 PENG for 0.187% filler loading for repeated (c) finger tapping, (d) mechanical impact (~105 kPa) and (e) bend/release action, (f) long-term stability performance of the PVDF-WS2 PENG in laboratory atmosphere for 12 weeks on repeated mechanical tapping and bending.

*Source:* Reprinted with permission from Ref. [138]. Copyright © 2021. The Royal Society of Chemistry.

### 3.2.7 APPLICATION OF GREEN POLYMERIC NANOMATERIALS FOR ENERGY APPLICATIONS

Green PNCs are very important as they possess some unique and special properties that combine the advantages of natural or synthetic nanofiller and organic polymers. The combination of green chemistry techniques with nanotechnology has vital importance in future nanotechnology. Owing to the rapid fossil fuel usage and serious ecological concerns, options for green energy development and its storage technology have become an emergent area of research. Integrating green NCs with active materials forms a flexible film/aerogel/3D structure as a substrate for powering portable electronics such as solar cells, piezoelectric materials, and LIBs, and several renewable energies (wind blades), sustainable energy (solar/fuel cell) environmental pollution prevention and medical applications. Research shows that the power transduction ability of solar cells and the piezoelectric performance of materials can be enhanced when the matrix is incorporated with nanomaterials such as nanocellulose.

Carbon nanomaterial composites are widely used in energy storage devices, and owing to nano-size and delocalized π electrons, charge transfer mobility has been enhanced several times and also improves the stability and efficiency of energy devices.[139] Nanocarbon serves to enhance thermal stability, dimensional resistance, and proton exchange performance by providing efficient charge transport at polymer/ nanocarbon surfaces. Enhancement in the photocurrent over the visible and near ultraviolet energy range can be obtained from MWCNTs when coated with copper NPs. MWCNTs are somewhat easier to process than SWCNTs because they are less prone to making tight bundles. Due to the presence of various consecutive cylindrical tubes, Multiwalled CNTs should be even more suitable for achieving charge transfer and transport than SWCNTs.[140]

Paper made from nanocellulose could be highly transparent with significant scattering along the light transport direction[141] and proved that nanocellulose paper could be a supreme substrate for supporting solar cells. Zhou et al. demonstrated organic solar cells on cellulose nanocrystal-CNC substrates, which yielded higher efficiencies.[142] Fang et al. enhanced the nanocellulose paper using a TEMPO-oxidized cellulose as the building unit and exhibited 96% optical transparency.[143] Kim et al.[144] regenerated bacteria cellulose to obtain flexible, uniform, and transparent cellulose film, and when these films were assembled with Cu to form a curved cellulose it showed a higher output. Guo et al. used the advantage of the morphological sequence of filter papers made from natural cellulose and subsequent calcination process to prepare nanotubular hollow SiOx.[145]

## 3.3 CONCLUSIONS AND FUTURE PERSPECTIVES

PNCs utilized for energy storage, energy saving, and energy harvesting applications have been reviewed here. The remarkable progress in processing, characterization, and designing of nanomaterials has opened considerable opportunities for engineering properties of the PNC systems. In the recent past, a new array of materials, NCs, have become of interest owing to the probability of increasing their useful properties. The design of PNCs requires the rational selection of NPs and polymer matrices. By wisely selecting a combination of polymer matrix, NPs, and fillers, the properties can be tuned, and the energy density has been markedly improved in the PNCs.

However, there are still many problems that remain unaddressed, like the limited availability of high-quality nanofillers, their increased production cost, innate difficulty to scale up, and lack of an in-depth understanding and predictive capability of the key process-structure-property relationships. Meanwhile, from the perspective of sustainability, more and more research efforts are needed on novel PNCs from clean, sustainable, and renewable energy sources. The future direction of PNCs as smart material for energy application is bright and is much awaited to become one of the supreme catalysts for the advancement of highly potent and innovative energy devices, especially in achieving possible renewable energy solutions.

## KEYWORDS

- **polymer nanocomposites**
- **photovoltaic cell**
- **supercapacitors**
- **fuel cell**
- **hydrogen storage**
- **piezoelectric polymer nanocomposites**

## REFERENCES


1. Fu, S.; Sun, Z.; Huang, P.; Li, Y.; Hu, N. Some Basic Aspects of Polymer Nanocomposites: A Critical Review. *Nano Mater. Sci.* **2019,** *1* (1), 2–30.
2. Fawaz, J.; Mittal, V. Synthesis of Polymer Nanocomposites: Review of Various Techniques. *Synth. Techniq. Polym. Nanocompos.* **2014,** 1–30.
3. Bergman, J. S.; Coates, G. W.; Chen, H.; Giannelis, E. P.; Thomas, M. G. Synthesis and Characterization of Polyolefin–Silicate Nanocomposites: A Catalyst Intercalation and In Situ Polymerization Approach. *Chem. Commun.* **1999,** *21*, 2179–2180.
4. Gou, J.; Zhuge, J.; Liang, F. Processing of Polymer Nanocomposites. In: *Manufacturing Techniques for Polymer Matrix Composites (PMCs)*; Woodhead Publishing, 2012; pp 95–119.
5. Bokobza, L. Spectroscopic Techniques for the Characterization of Polymer Nanocomposites: A Review. *Polymers* **2017,** *10* (1), 7.
6. Szymoniak, P.; Qu, X.; Schönhals, A.; Sturm, H. Characterization of Polymer Nanocomposites. In: *Acting Principles of Nano-Scaled Matrix Additives for Composite Structures*; Springer International Publishing: Cham, 2021; pp 55–77.
7. Tjong, S. C. Structural and Mechanical Properties of Polymer Nanocomposites. *Mater. Sci. Eng.: R Rep.* **2006,** *53* (3–4), 73–197.
8. Psarras, G. C. Conductivity and Dielectric Characterization of Polymer Nanocomposites. In: *Physical Properties and Applications of Polymer Nanocomposites*; Woodhead Publishing, 2010; pp 31–69.
9. Mittal, V., Ed. *Characterization Techniques for Polymer Nanocomposites*; John Wiley & Sons, 2012.
10. Amin, K. F.; Sen, A.; Hoque, M. E. Polymer Nanocomposites for Energy. In: *Advanced Polymer Nanocomposites*; Woodhead Publishing, 2022; pp 335–372.
11. Dargusch, M.; Liu, W.; Chen, Z. Thermoelectric Generators: Alternative Power Supply for Wearable Electrocardiographic Systems. *Adv. Sci.* **2020,** *7* (18), 2001362. https://doi.org/10.1002/advs.202001362.
12. Muchuweni, E.; Mombeshora, E. T. Enhanced Thermoelectric Performance by Single-Walled Carbon Nanotube Composites for Thermoelectric Generators: A Review. *Appl. Surf. Sci. Adv.* **2023,** *13*, 100379.
13. Sarkar, K.; Debnath, A.; Deb, K.; Bera, A.; Saha, B. Effect of NiO Incorporation in Charge Transport of Polyaniline: Improved Polymer Based Thermoelectric Generator. *Energy* **2019,** *177*, 203–210.
14. Debnath, A.; Deb, K.; Sarkar, K.; Saha, B. Improved Thermoelectric Performance in $TiO_2$ Incorporated Polyaniline: A Polymer-Based Hybrid Material for Thermoelectric Generators. *J. Electron. Mater.* **2020,** *49*, 5028–5036.
15. Du, Y.; Cai, K. F.; Chen, S.; Cizek, P.; Lin, T. Facile Preparation and Thermoelectric Properties of $Bi_2Te_3$ Based Alloy Nanosheet/PEDOT: PSS Composite Films. *ACS Appl. Mater. Interfaces* **2014,** *6* (8), 5735–5743.
16. Palaporn, D.; Mongkolthanaruk, W.; Faungnawakij, K.; Kurosaki, K.; Pinitsoontorn, S. Flexible Thermoelectric Paper and Its Thermoelectric Generator from Bacterial Cellulose/$Ag_2Se$ Nanocomposites. *ACS Appl. Energy Mater.* **2022,** *5* (3), 3489–3501.
17. Hou, W.; Xiao, Y.; Han, G.; Lin, J. Y. The Applications of Polymers in Solar Cells: A Review. *Polymers* **2019,** *11* (1), 143.

18. Di Mare, S.; Menossi, D.; Salavei, A.; Artegiani, E.; Piccinelli, F.; Kumar, A.; Mariotto, G.; Romeo, A. SnS Thin Film Solar Cells: Perspectives and Limitations. *Coatings* **2017,** *7* (2), 34.
19. Alhamada, T. F.; Hanim, M. A.; Saidur, R.; Nuraini, A.; Hasan, W. W.; Jung, D. W. Recent Advances in Polymer and Perovskite Based Third-Generation Solar Cell Devices. *Mater. Today: Proc.* **2022**.
20. Ahmad, Z.; Farooq, E.; Nazar, R.; Mehmood, U.; Fareed, I. Development of Multi-Walled Carbon Nanotube/Polythiophene (MWCNT/PTh) Nanocomposites for Platinum-Free Dye-Sensitized Solar Cells (DSSCs). *Solar Energy* **2022,** *245*, 153–157.
21. Rafique, S.; Rashid, I.; Sharif, R. Cost Effective Dye Sensitized Solar Cell Based on Novel Cu Polypyrrole Multiwall Carbon Nanotubes Nanocomposites Counter Electrode. *Sci. Rep.* **2021,** *11* (1), 14830.
22. Wu, J.; Lan, Z.; Lin, J.; Huang, M.; Huang, Y.; Fan, L.; Luo, G. Electrolytes in Dye-Sensitized Solar Cells. *Chem. Rev.* **2015,** *115* (5), 2136–2173.
23. Vinoth, S.; Kanimozhi, G.; Narsimulu, D.; Kumar, H.; Srinadhu, E. S.; Satyanarayana, N. Ionic Relaxation of Electrospun Nanocomposite Polymer-Blend Quasi-Solid Electrolyte for High Photovoltaic Performance of Dye-Sensitized Solar Cells. *Mater. Chem. Phys.* **2020,** *250*, 122945.
24. Sultana, M.; Mehmood, U.; Nazar, R.; Gill, Y. Q. Development of Multiwalled Carbon Nanotubes/Polyaniline Nanocomposites Based Electrolyte for Quasi-Solid-State Dye-Sensitized Solar Cells. *Int. J. Energy Res.* **2022,** *46* (7), 9911–9918.
25. Wen, J.; Sun, Z.; Qiao, Y.; Zhou, Y.; Liu, Y.; Zhang, Q.; Liu, Y.; Jiao, S. $Ti_3C_2$ MXene-Reduced Graphene Oxide Composite Polymer-Based Printable Electrolyte for Quasi-Solid-State Dye-Sensitized Solar Cells. *ACS Appl. Energy Mater.* **2022,** *5* (3), 3329–3338.
26. Radaoui, M.; Fredj, A. B.; Romdhane, S.; Bouguerra, N.; Egbe, D. A.; Bouchriha, H. New Conjugated Polymer/Fullerene Nanocomposite for Energy Storage and Organic Solar Cell Devices: Studies of the Impedance Spectroscopy and Dielectric Properties. *Synth. Metals* **2022,** *283*, 116987.
27. Dou, Q.; Whatley, T.; Syed, T.; Wei, W.; Wang, H. Carbon Nanomaterials–Polymer Composites for Perovskite Solar Cells: Preparation, Properties and Applications. *J. Mater. Chem. A.* **2022,** *10* (37), 19211–19230.
28. Rajamanickam, N.; Kumari, S.; Vendra, V. K.; Lavery, B. W.; Spurgeon, J.; Druffel, T.; Sunkara, M. K. Stable and Durable $CH_3NH_3PbI_3$ Perovskite Solar Cells at Ambient Conditions. *Nanotechnology* **2016,** *27* (23), 235404.
29. Lou, Q.; Lou, G.; Peng, R.; Liu, Z.; Wang, W.; Ji, M.; Chen, C.; Zhang, X.; Liu, C.; Ge, Z. Synergistic Effect of Lewis Base Polymers and Graphene in Enhancing the Efficiency of Perovskite Solar Cells. *ACS Appl. Energy Mater.* **2021,** *4* (4), 3928–3936.
30. Zhou, Y.; Yin, X.; Luo, Q.; Zhao, X.; Zhou, D.; Han, J.; Hao, F.; Tai, M.; Li, J.; Liu, P.; Jiang, K. Efficiently Improving the Stability of Inverted Perovskite Solar Cells by Employing Polyethylenimine-Modified Carbon Nanotubes as Electrodes. *ACS Appl. Mater. Interfaces* **2018,** *10* (37), 31384–31393.
31. Jang, J. H.; Kim, B. J.; Kim, J. H.; Han, E.; Choi, E. Y.; Ji, C. H.; Kim, K. T.; Kim, J.; Park, N. A Novel Approach for the Development of Moisture Encapsulation Poly(vinyl alcohol-co-ethylene) for Perovskite Solar Cells. *ACS Omega* **2019,** *4* (5), 9211–9218.

32. Mohseni, H. R.; Dehghanipour, M.; Dehghan, N.; Tamaddon, F.; Ahmadi, M.; Sabet, M.; Behjat, A. J. Enhancement of the Photovoltaic Performance and the Stability of Perovskite Solar Cells via the Modification of Electron Transport Layers with Reduced Graphene Oxide/Polyaniline Composite. *Solar Energy* **2021,** *213*, 59–66.
33. Yu, J. C.; Hong, J. A.; Jung, E. D.; Kim, D. B.; Baek, S. M.; Lee, S.; Cho, S.; Park, S. S.; Choi, K. J.; Song, M. H. Highly Efficient and Stable Inverted Perovskite Solar Cell Employing PEDOT: GO Composite Layer as a Hole Transport Layer. *Sci. Rep.* **2018,** *8* (1), 1–9.
34. Cho, S. P.; Han, G.; Seo, Y. H.; Noh, Y. J.; Sohn, J. Y.; Hwang, I. T.; Shin, J.; Jung, C. H.; Na, S. I. Highly-Dispersible Reduced Graphene Oxide/Polymer Nanocomposites as Efficient Hole-Transporting Materials for Perovskite Solar Cells. *Compos. Sci. Technol.* **2021,** *201*, 108548.
35. Subramanian, V.; Hari Prasad, K.; Das, H. T.; Ganapathy, K.; Nallani, S.; Maiyalagan, T. Novel Dispersion of 1D Nanofiber Fillers for Fast Ion-Conducting Nanocomposite Polymer Blend Quasi-Solid Electrolytes for Dye-Sensitized Solar Cells. *ACS Omega* **2022,** *7* (2), 1658–1670.
36. Kumar, S.; Manikandan, V. S.; Palai, A. K.; Mohanty, S.; Nayak, S. K. $Fe_2O_3$ as an Efficient Filler in PVDF-HFP Based Polymeric Electrolyte for Dye Sensitized Solar Cell Application. *Solid State Ion.* **2019,** *332*, 10–15.
37. Thomas, M.; Rajiv, S. Dye-sensitized Solar Cells Based on an Electrospun Polymer Nanocomposite Membrane as Electrolyte. *New J. Chem.* **2019,** *43* (11), 4444–4454.
38. Zebardastan, N.; Ramesh, S.; Ramesh, K. Performance Enhancement of Poly(vinylidene fluoride-co-hexafluoro propylene)/Polyethylene Oxide Based Nanocomposite Polymer Electrolyte with ZnO Nanofiller for Dye-Sensitized Solar Cell. *Organ. Electron.* **2017,** *49*, 292–299.
39. Liow, K. S.; Sipaut, C. S.; Jafarzadeh, M. Polypyrrole-and Polyaniline-Surface Modified Nanosilica as Quasi-Solid State Electrolyte Ingredients for Dye-Sensitized Solar Cells. *J. Mater. Sci.: Mater. Electron.*. **2018,** *29*, 21097–21108.
40. Saidi, N. M.; Omar, F. S.; Numan, A.; Apperley, D. C.; Algaradah, M. M.; Kasi, R.; Avestro, A. J.; Subramaniam, R. T. Enhancing the Efficiency of a Dye-Sensitized Solar Cell Based on a Metal Oxide Nanocomposite Gel Polymer Electrolyte. *ACS Appl. Mater. Interfaces* **2019,** *11* (33), 30185–30196.
41. Ike, J. N.; Adedeji, M. A.; Mola, G. T. Silver Doped ZnS Core-Shell Nanocomposites to Promote Photon Harvesting in Polymer Cells. *Phys. B: Condens. Matter*. **2022,** *646*, 414347.
42. Arbab, E. A.; Mola, G. T. Metals Decorated Nanocomposite Assisted Charge Transport in Polymer Solar Cell. *Mater. Sci. Semiconductor Process*. **2019,** *91*, 1–8.
43. Shabzendedar, S.; Modarresi-Alam, A. R.; Noroozifar, M.; Kerman, K. Core-Shell Nanocomposite of Superparamagnetic $Fe_3O_4$ Nanoparticles with Poly(m-aminobenzenesulfonic acid) for Polymer Solar Cells. *Organ. Electron.* **2020,** *77*, 105462.
44. Shabzendedar, S.; Modarresi-Alam, A. R.; Paymozd, F.; Yarmohamadi-Vasel, M.; Kaedi, F.; Li, Y. Solar Cells Containing Two Novel Superparamagnetic Nanocomposites of $Fe_3O_4$-$TiO_2$-Poly(m-aminobenzenesulfonic acid). *Synth. Metals* **2023,** *295*, 117335.
45. Ferrag, C.; Noroozifar, M.; Modarresi-Alam, A. R.; Kerman, K. Encapsulation of Poly(m-aminobenzodioxol)-$Fe_3O_4$ Superparamagnetic Nanorods and Iron (III)

Thiocyanate Complex in Hydrogel Toward Hybrid Solar Cells. *J. Environ. Chem. Eng.* **2021,** *9* (4), 105612.
46. Golshani, Z.; Maghsoudi, S.; Hosseini, S. M. Enhancement of the Photovoltaic Performance of HTL-Free-Perovskite Solar Cells Based on Carbon Electrode via the Modification of Electron Transport Layer with Copper Oxide@ Polyaniline Nanocomposite. *Energy Rep.* **2022,** *8*, 13596–13609.
47. Idumah, C. I.; Ezeani, E. O.; Nwuzor, I. C. A Review: Advancements in Conductive Polymers Nanocomposites. *Polym.-Plastics Technol. Mater.* **2021,** *60* (7), 756–783.
48. Sharma, S.; Sudhakara, P.; Omran, A. A.; Singh, J.; Ilyas, R. A. Recent Trends and Developments in Conducting Polymer Nanocomposites for Multifunctional Applications. *Polymers* **2021,** *13* (17), 2898.
49. Ahmed, U.; Shahid, M. M.; Shahabuddin, S.; Abd Rahim, N.; Alizadeh, M.; Pandey, A. K.; Sagadevan, S. An Efficient Platform Based on Strontium Titanate Nanocubes Interleaved Polypyrrole Nanohybrid as Counter Electrode for Dye-Sensitized Solar Cell. *J. Alloys Compd.* **2021,** *860*, 158228.
50. He, J.; Duffy, N. W.; Pringle, J. M.; Cheng, Y. B. Conducting Polymer and Titanium Carbide-Based Nanocomposites as Efficient Counter Electrodes for Dye-Sensitized Solar Cells. *Electrochim. Acta* **2013,** *105*, 275–281.
51. Zoromba, M. S.; Abdel-Aziz, M. H.; Bassyouni, M.; Bahaitham, H.; Al-Hossainy, A. F. Poly(o-phenylenediamine) Thin Film for Organic Solar Cell Applications. *J. Solid State Electrochem.* **2018,** *22*, 3673–3687.
52. Al-Hossainy, A. F.; Thabet, H. K.; Zoromba, M. S.; Ibrahim, A. Facile Synthesis and Fabrication of a Poly(ortho-anthranilic acid) Emeraldine Salt Thin Film for Solar Cell Applications. *New J. Chem.* **2018,** *42* (12), 10386–10395.
53. Al-Hossainy, A. F.; Zoromba, M. S.; Abdel-Aziz, M. H.; Bassyouni, M.; Attar, A.; Zwawi, M.; Abd-Elmageed, A. A.; Maddah, H. A.; Slimane, A. B. Fabrication of Heterojunction Diode Using Doped-Poly(ortho-aminophenol) for Solar Cells Applications. *Phys. B: Condens. Matter.* **2019,** *566*, 6–16.
54. Abdel-Aziz, M. H.; Maddah, H. A.; Zoromba, M. S.; Al-Hossainy, A. F. One-Dimensional Ternary Conducting Polymers Blend with 9.26% Power Conversion Efficiency for Photovoltaic Devices Applications. *Alexandria Eng. J.* **2023,** *66*, 475–488.
55. Adimule, V.; Suryavanshi, A.; BC Y, Nandi, S. S. A Facile Synthesis of Poly (3-octyl thiophene), $Ni_{0.4}Sr_{0.6}TiO_3$ Hybrid Nanocomposites for Solar Cell Applications. *Macromol. Symposia* **2020,** *392* (1), 2000001.
56. Yanik, M. O.; Yigit, E. A.; Akansu, Y. E.; Sahmetlioglu, E. Magnetic Conductive Polymer-Graphene Nanocomposites Based Supercapacitors for Energy Storage. *Energy* **2017,** *138*, 883–889.
57. Shen, F.; Pankratov, D.; Chi, Q. Graphene-Conducting Polymer Nanocomposites for Enhancing Electrochemical Capacitive Energy Storage. *Curr. Opin. Electrochem.* **2017,** *4* (1), 133–144.
58. Zhang, X.; Samorì, P. Graphene/Polymer Nanocomposites for Supercapacitors. *ChemNanoMat* **2017,** *3* (6), 362–372.
59. Ferrari, S.; Nair, J. R.; Zhou, Y.; Wan, C. Polymer Nanocomposites for Lithium Battery Applications. In: *Polymer-Based Nanocomposites for Energy and Environmental Applications*; Woodhead Publishing, 2018; pp 283–313.

60. Prateek, Thakur, V. K.; Gupta, R. K. Recent Progress on Ferroelectric Polymer-Based Nanocomposites for High Energy Density Capacitors: Synthesis, Dielectric Properties, and Future Aspects. *Chem. Rev.* **2016,** *116* (7), 4260–4317.
61. He, B. L.; Dong, B.; Wang, W.; Li, H. L. Performance of Polyaniline/Multi-Walled Carbon Nanotubes Composites as Cathode for Rechargeable Lithium Batteries. *Mater. Chem. Phys.* **2009,** *114* (1), 371–375.
62. Zhang, X.; Samorì, P. Graphene/Polymer Nanocomposites for Supercapacitors. *ChemNanoMat* **2017,** *3* (6), 362–732.
63. Sydulu Singu, B.; Srinivasan, P.; Yoon, K. R. Emulsion Polymerization Method for Polyaniline-Multiwalled Carbon Nanotube Nanocomposites as Supercapacitor Materials. *J. Solid State Electrochem.* **2016,** *20*, 3447–3457.
64. Gopal, J.; Muthu, M.; Sivanesan, I. A Comprehensive Compilation of Graphene/Fullerene Polymer Nanocomposites for Electrochemical Energy Storage. *Polymers* **2023,** *15* (3), 701.
65. Cai, C.; Wang, Y. Novel Nanocomposite Materials for Advanced Li-Ion Rechargeable Batteries. *Materials* **2009,** *2* (3), 1205–1238.
66. Zhang, F.; Cao, H.; Yue, D.; Zhang, J.; Qu, M. Enhanced Anode Performances of Polyaniline–$TiO_2$–Reduced Graphene Oxide Nanocomposites for Lithium Ion Batteries. *Inorgan. Chem.* **2012,** *51* (17), 9544–9551.
67. Zhu, W. C.; He, P. Q.; Tien, H. C.; Liu, H. L.; Chen, W. C.; Lv, W.; Lee, W. Y. Solvent-Enhanced Transparent Stretchable Polymer Nanocomposite Electrode for Supercapacitors. *ACS Appl. Energy Mater.* **2021,** *4* (3), 2266–2274.
68. Ansari, S. A.; Parveen, N.; Han, T. H.; Ansari, M. O.; Cho, M. H. Fibrous Polyaniline@ Manganese Oxide Nanocomposites as Supercapacitor Electrode Materials and Cathode Catalysts for Improved Power Production in Microbial Fuel Cells. *Phys. Chem. Chem. Phys.* **2016,** *18* (13), 9053–9060.
69. Singu, B. S.; Palaniappan, S.; Yoon, K. R. Polyaniline–Nickel Oxide Nanocomposites for Supercapacitor. *J. Appl. Electrochem.* **2016,** *46*, 1039–1047.
70. Shown, I.; Ganguly, A.; Chen, L. C.; Chen, K. H. Conducting Polymer-Based Flexible Supercapacitor. *Energy Sci. Eng.* **2015,** *3* (1), 2–6.
71. Alvi, F.; Ram, M. K.; Basnayaka, P. A.; Stefanakos, E.; Goswami, Y.; Kumar, A. Graphene–Polyethylenedioxythiophene Conducting Polymer Nanocomposite Based Supercapacitor. *Electrochim. Acta.* **2011,** *56* (25), 9406–9412.
72. Ramaprabhu, S. Poly (p-phenylenediamine)/Graphene Nanocomposites for Supercapacitor Applications. *J. Mater. Chem.* **2012,** *22* (36), 18775–18783.
73. Chen, X.; Zhao, Y.; Li, L.; Wang, Y.; Wang, J.; Xiong, J.; Du, S.; Zhang, P.; Shi, X.; Yu, J. MXene/Polymer Nanocomposites: Preparation, Properties, and Applications. *Polym. Rev.* **2021,** *61* (1), 80–115.
74. Adekoya, G. J.; Adekoya, O. C.; Sadiku, R. E.; Hamam, Y.; Ray, S. S. Applications of MXene-Containing Polypyrrole Nanocomposites in Electrochemical Energy Storage and Conversion. *ACS Omega.* **2022,** *7* (44), 39498–39519.
75. Bharti, B.; Kumar, R.; Kumar, H.; Li, H.; Zha, X.; Ouyang, F. Advanced Applications and Current Status of Green Nanotechnology in the Environmental Industry. In: *Green Functionalized Nanomaterials for Environmental Applications*; Elsevier, 2022; pp 303–340

76. Kim, D. J.; Jo, M. J.; Nam, S. Y. A Review of Polymer–Nanocomposite Electrolyte Membranes for Fuel Cell Application. *J. Ind. Eng. Chem.* **2015,** *21*, 36–52.
77. Peighambardoust, S. J.; Rowshanzamir, S.; Amjadi, M. Review of the Proton Exchange Membranes for Fuel Cell Applications. *Int. J. Hydrogen Energy*. **2010,** *35* (17), 9349–9384.
78. Abouzari-Lotf, E.; Etesami, M.; Nasef, M. M. Carbon-Based Nanocomposite Proton Exchange Membranes for Fuel Cells. In: *Carbon-Based Polymer Nanocomposites for Environmental and Energy Applications*; Elsevier, 2018; pp 437–461.
79. Baker, A. M.; Wang, L.; Johnson, W. B.; Prasad, A. K.; Advani, S. G. Nafion Membranes Reinforced with Ceria-Coated Multiwall Carbon Nanotubes for Improved Mechanical and Chemical Durability in Polymer Electrolyte Membrane Fuel Cells. *J. Phys. Chem. C*. **2014,** *118* (46), 26796–26802.
80. Jafary, T.; Ghasemi, M.; Alam, J.; Aljlil, S. A.; Yusup, S. Carbon-Based Polymer Nanocomposites as Electrodes for Microbial Fuel Cells. In: *Carbon-Based Polymer Nanocomposites for Environmental and Energy Applications*; Elsevier, 2018; pp 361–390.
81. Yang, W.; Peng, Y.; Zhang, Y.; Lu, J. E.; Li, J.; Chen, S. Air Cathode Catalysts of Microbial Fuel Cell by Nitrogen-Doped Carbon Aerogels. *ACS Sustain. Chem. Eng.* **2018,** *7* (4), 3917–3924.
82. Liang, P.; Wang, H.; Xia, X.; Huang, X.; Mo, Y.; Cao, X.; Fan, M. Carbon Nanotube Powders as Electrode Modifier to Enhance the Activity of Anodic Biofilm in Microbial Fuel Cells. *Biosens. Bioelectron.* **2011,** *26* (6), 3000–3004.
83. Lin, J. S.; Kumar, S. R.; Ma, W. T.; Shih, C. M.; Teng, L. W.; Yang, C. C.; Lue, S. J. Gradiently Distributed Iron Oxide@ Graphene Oxide Nanofillers in Quaternized Polyvinyl Alcohol Composite to Enhance Alkaline Fuel Cell Power Density. *J. Membr. Sci.* **2017,** *543*, 28–39.
84. Dessie, Y.; Tadesse, S.; Eswaramoorthy, R. Review on Manganese Oxide Based Biocatalyst in Microbial Fuel Cell: Nanocomposite Approach. *Mater. Sci. Energy Technol.* **2020,** *3*, 136–149.
85. Hindatu, Y.; Annuar, M. S.; Gumel, A. M. Mini-Review: Anode Modification for Improved Performance of Microbial Fuel Cell. *Renew. Sustain. Energy Rev.* **2017,** *73*, 236–248.
86. Nourbakhsh, F.; Mohsennia, M.; Pazouki, M. Highly Efficient Cathode for the Microbial Fuel Cell Using $LaXO_3$ (X=[Co, Mn, $Co_{0.5}Mn_{0.5}$]) Perovskite Nanoparticles as Electrocatalysts. *SN Appl. Sci.* **2020,** *2*, 1–9.
87. Yogarathinam, L. T.; Jaafar, J.; Ismail, A. F.; Junoh, H.; Samavati, A.; Goh, P. S.; Gangasalam, A.; Peter, J. Synthesis and Characterization of Conductive Polymer Coated Graphitic Carbon Nitride Embedded Sulfonated Poly(ether ether ketone) Membranes for Direct Methanol Fuel Cell Applications. *Int. J. Energy Res.* **2021,** *45* (11), 16649–16666.
88. Gahlot, S.; Sharma, P. P.; Kulshrestha, V.; Jha, P. K. SGO/SPES-Based Highly Conducting Polymer Electrolyte Membranes for Fuel Cell Application. *ACS Appl. Mater. Interfaces* **2014,** *6* (8), 5595–5601.
89. Muthumeenal, A.; Saraswathi, M. S.; Rana, D.; Nagendran, A. Fabrication and Electrochemical Properties of Highly Selective SPES/GO Composite Membranes for Direct Methanol Fuel Cells. *J. Environ. Chem. Eng.* **2017,** *5* (4), 3828–3833.

90. Elakkiya, S.; Arthanareeswaran, G.; Ismail, A. F.; Das, D. B.; Suganya, R. Polyaniline Coated Sulfonated $TiO_2$ Nanoparticles for Effective Application in Proton Conductive Polymer Membrane Fuel Cell. *Eur. Polym. J.* **2019,** *112*, 696–703.
91. Zhang, R. C.; Sun, D.; Zhang, R.; Lin, W. F.; Macias-Montero, M.; Patel, J.; Askari, S.; McDonald, C.; Mariotti, D.; Maguire, P. Gold Nanoparticle-Polymer Nanocomposites Synthesized by Room Temperature Atmospheric Pressure Plasma and Their Potential for Fuel Cell Electrocatalytic Application. *Sci. Rep.* **2017,** *7* (1), 1–9.
92. Elancheziyan, M.; Eswaran, M.; Shuck, C. E.; Senthilkumar, S.; Elumalai, S.; Dhanusuraman, R.; Ponnusamy, V. K. Facile Synthesis of Polyaniline/Titanium Carbide (MXene) Nanosheets/Palladium Nanocomposite for Efficient Electrocatalytic Oxidation of Methanol for Fuel Cell Application. *Fuel* **2021,** *303*, 121329.
93. Karatepe, N.; Yuca, N.; Şenkal, B. F. Synthesis of Carbon-Based Nano Materials for Hydrogen Storage. *Fullerenes Nanotubes Carbon Nanostruct.* **2013,** *21* (1), 31–46.
94. Silambarasan, D.; Vasu, V.; Iyakutti, K. Water Soluble Polymer-SWCNT-Based Composite for Hydrogen Storage. *IEEE Trans. Nanotechnol.* **2014,** *13* (2), 261–267.
95. Oliveira, A. D.; Beatrice, C. A.; Passador, F. R.; Pessan, L. A. Polyetherimide-Based Nanocomposites Materials For Hydrogen Storage. *AIP Conf. Proc* **2016,** *1779* (1), 040006.
96. Jeon, K. J.; Moon, H. R.; Ruminski, A. M.; Jiang, B.; Kisielowski, C.; Bardhan, R.; Urban, J. J. Air-Stable Magnesium Nanocomposites Provide Rapid and High-Capacity Hydrogen Storage Without Using Heavy-Metal Catalysts. *Nat. Mater.* **2011,** *10* (4), 286–290.
97. Makridis, S. S.; Gkanas, E. I.; Panagakos, G.; Kikkinides, E. S.; Stubos, A. K.; Wagener, P.; Barcikowski, S. Polymer-Stable Magnesium Nanocomposites Prepared by Laser Ablation for Efficient Hydrogen Storage. *Int. J. Hydrogen Energy* **2013,** *38* (26), 11530–11535.
98. Ruminski, A. M.; Bardhan, R.; Brand, A.; Aloni, S.; Urban, J. J. Synergistic Enhancement of Hydrogen Storage and Air Stability via Mg Nanocrystal–Polymer Interfacial Interactions. *Energy Environ. Sci.* **2013,** *6* (11), 3267–3271.
99. Beatrice, C. A.; Oliveira, A. D.; Passador, F. R.; Pessan, L. A. Sulfonated Poly(ether imide)/Aluminium Nanocomposites for Hydrogen Storage. *AIP Conf. Proc.* **2016,** *1779* (1), 040004.
100. de Almeida Neto, G. R.; Beatrice, C. A.; Leiva, D. R.; Pessan, L. A. Polyetherimide-$LaNi_5$ Composite Films for Hydrogen Storage Applications. *Int. J. Hydrogen Energy* **2021,** *46* (46), 23767–23778.
101. Kim, B. H.; Hong, W. G.; Lee, S. M.; Yun, Y. J.; Yu, H. Y.; Oh, S. Y.; Kim, C. H.; Kim, Y. Y.; Kim H. J. Enhancement of Hydrogen Storage Capacity in Polyaniline-Vanadium Pentoxide Nanocomposites. *Int. J. Hydrogen Energy* **2010,** *35* (3), 1300–1304.
102. Stejskal, J.; Gilbert, R. G. Polyaniline Preparation of a Conducting Polymer (IUPAC Technical Report). *Pure Appl. Chem.* **2002,** *74* (5), 857–867.
103. Beatrice, C. A.; Moreira, B. R.; de Oliveira, A. D.; Passador, F. R.; de Almeida Neto, G. R.; Leiva, D. R.; Pessan, L. A. Development of Polymer Nanocomposites with Sodium Alanate for Hydrogen Storage. *Int. J. Hydrogen Energy* **2020,** *45* (8), 5337–5346.
104. Srinivasan, S. S.; Ratnadurai, R.; Niemann, M. U.; Phani, A. R.; Goswami, D. Y.; Stefanakos, E. K. Reversible Hydrogen Storage in Electrospun Polyaniline Fibers. *Int. J. Hydrogen Energy* **2010,** *35* (1), 225–230.

105. Attia, N. F.; Geckeler, K. E. Polyaniline–Polypyrrole Composites with Enhanced Hydrogen Storage Capacities. *Macromol. Rapid Commun.* **2013,** *34* (11), 931–937.
106. Attia, N. F.; Menemparabath, M. M.; Arepalli, S.; Geckeler, K. E. Inorganic Nanotube Composites Based on Polyaniline: Potential Room-Temperature Hydrogen Storage Materials. *Int. J. Hydrogen Energy* **2013,** *38* (22), 9251–9262.
107. Muthu, R. N.; Rajashabala, S.; Kannan, R. Synthesis and Characterization of Polymer (Sulfonated Poly-Ether-Ether-Ketone) Based Nanocomposite (H-Boron Nitride) Membrane for Hydrogen Storage. *Int. J. Hydrogen Energy* **2015,** *40* (4), 1836–1845.
108. Prateek, Thakur, V. K.; Gupta, R. K. Recent Progress on Ferroelectric Polymer-Based Nanocomposites for High Energy Density Capacitors: Synthesis, Dielectric Properties, and Future Aspects. *Chem. Rev.* **2016,** *116* (7), 4260–4317.
109. Ponnamma, D.; Erturk, A.; Parangusan, H.; Deshmukh, K.; Ahamed, M. B.; Al Ali Al-Maadeed, M. Stretchable Quaternary Phasic PVDF-HFP Nanocomposite Films Containing Graphene-Titania-$SrTiO_3$ for Mechanical Energy Harvesting. *Emergent Mater.* **2018,** *1*, 55–65.
110. Surmenev, R. A.; Orlova, T.; Chernozem, R. V.; Ivanova, A. A.; Bartasyte, A.; Mathur, S.; Surmeneva, M. A. Hybrid Lead-Free Polymer-Based Nanocomposites with Improved Piezoelectric Response for Biomedical Energy-Harvesting Applications: A Review. *Nano Energy* **2019,** *62*, 475–506.
111. Zhang, J.; Fang, Z.; Shu, C.; Zhang, J.; Zhang, Q.; Li, C. A Rotational Piezoelectric Energy Harvester for Efficient Wind Energy Harvesting. *Sens. Actuat. A: Phys.* **2017,** *262*, 123–129.
112. Lotfian, S.; Giraudmaillet, C.; Yoosefinejad, A.; Thakur, V. K.; Nezhad, H. Y. Electrospun Piezoelectric Polymer Nanofiber Layers for Enabling In Situ Measurement in High-Performance Composite Laminates. *ACS Omega* **2018,** *3* (8), 8891–8902.
113. Bhunia, R.; Das, S.; Dalui, S.; Hussain, S.; Paul, R.; Bhar, R.; Pal, A. K. Flexible Nano-ZnO/Polyvinylidene Difluoride Piezoelectric Composite Films as Energy Harvester. *Appl. Phys. A* **2016,** *122*, 1–3.
114. Shin, S. H.; Kim, Y. H.; Lee, M. H.; Jung, J. Y.; Nah, J. *ACS Nano* **2014,** *8*, 2766.
115. Yaqoob, U.; Uddin, A. I.; Chung, G. S. A Novel Tri-Layer Flexible Piezoelectric Nanogenerator Based on Surface-Modified Graphene and PVDF-$BaTiO_3$ Nanocomposites. *Appl. Surf. Sci.* **2017,** *405*, 420–426.
116. Shetty, S.; Mahendran, A.; Anandhan, S. Development of a New Flexible Nanogenerator from Electrospun Nanofabric Based on PVDF/Talc Nanosheet Composites. *Soft Matter.* **2020,** *16* (24), 5679–5688.
117. Park, K. I.; Bae, S. B.; Yang, S. H.; Lee, H. I.; Lee, K.; Lee, S. J. Lead-Free $BaTiO_3$ Nanowires-Based Flexible Nanocomposite Generator. *Nanoscale* **2014,** *6* (15), 8962–8968.
118. Jung, J. H.; Lee, M.; Hong, J. I.; Ding, Y.; Chen, C. Y.; Chou, L. J.; Wang, Z. L. Lead-Free $NaNbO_3$ Nanowires for a High Output Piezoelectric Nanogenerator. *ACS Nano.* **2011,** *5* (12), 10041–10046.
119. Paik, H.; Choi, Y. Y.; Hong, S.; No, K. Effect of Ag Nanoparticle Concentration on the Electrical and Ferroelectric Properties of Ag/P (VDF-TrFE) Composite Films. *Sci. Rep.* **2015,** *5* (1), 13209.
120. Ahn, Y.; Lim, J. Y.; Hong, S. M.; Lee, J.; Ha, J.; Choi, H. J.; Seo, Y. Enhanced Piezoelectric Properties of Electrospun Poly(vinylidene fluoride)/Multiwalled

Carbon Nanotube Composites Due to High β-Phase Formation in Poly(vinylidene fluoride). *J. Phys. Chem. C* **2013,** *117* (22), 11791–11799.
121. Xue, J.; Wu, L.; Hu, N.; Qiu, J.; Chang, C.; Atobe, S.; Fukunaga, H.; Watanabe, T.; Liu, Y.; Ning, H.; Li, J. Evaluation of Piezoelectric Property of Reduced Graphene Oxide (rGO)–Poly (vinylidene fluoride) Nanocomposites. *Nanoscale* **2012,** *4* (22), 7250–7255.
122. Wu, L.; Yuan, W.; Hu, N.; Wang, Z.; Chen, C.; Qiu, J.; Ying, J.; Li, Y. Improved Piezoelectricity of PVDF-HFP/Carbon Black Composite Films. *J. Phys. D: Appl. Phys.* **2014,** *47* (13), 135302.
123. Pusty, M.; Sharma, A.; Sinha, L.; Chaudhary, A.; Shirage, P. Comparative Study with a Unique Arrangement to Tap Piezoelectric Output to Realize a Self Poled PVDF Based Nanocomposite for Energy Harvesting Applications. *ChemistrySelect* **2017,** *2* (9), 2774–2782.
124. Singh, H. H.; Singh, S.; Khare, N. Design of Flexible PVDF/NaNbO3/RGO Nanogenerator and Understanding the Role of Nanofillers in the Output Voltage Signal. *Compos. Sci. Technol.* **2017,** *149*, 127–133.
125. Yang, L.; Cheng, M.; Lyu, W.; Shen, M.; Qiu, J.; Ji, H.; Zhao, Q. Tunable Piezoelectric Performance of Flexible PVDF Based Nanocomposites from MWCNTs/Graphene/$MnO_2$ Three-Dimensional Architectures Under Low Poling Electric Fields. *Compos. Part A: Appl. Sci. Manuf.* **2018,** *107*, 536–544.
126. Ponnamma, D.; Erturk, A.; Parangusan, H.; Deshmukh, K.; Ahamed, M. B.; Al Ali Al-Maadeed, M. Stretchable Quaternary Phasic PVDF-HFP Nanocomposite Films Containing Graphene-Titania-$SrTiO_3$ for Mechanical Energy Harvesting. *Emergent Mater.* **2018,** *1*, 55–65.
127. Abbasipour, M.; Khajavi, R.; Yousefi, A. A.; Yazdanshenas, M. E.; Razaghian, F. The Piezoelectric Response of Electrospun PVDF Nanofibers with Graphene Oxide, Graphene, and Halloysite Nanofillers: A Comparative Study. *J. Mater. Sci.: Mater. Electron.* **2017,** *28*, 15942–15952.
128. Park, K.-I.; Lee, M.; Liu, Y.; Moon, S.; Hwang, G.-T.; Zhu, G.; Kim, J. E.; Kim, S. O.; Kim, D. K.; Wang, Z. L.; Lee, K. J. *Adv. Mater.* **2012,** *24*, 2999.
129. Park, K. I.; Jeong, C. K.; Ryu, J.; Hwang, G. T.; Lee, K. J. Flexible and Large-Area Nanocomposite Generators Based on Lead Zirconate Titanate Particles and Carbon Nanotubes. *Adv. Energy Mater.* **2013,** *3* (12), 1539–1544.
130. Pal, A.; Sasmal, A.; Manoj, B.; Rao, D. P.; Haldar, A. K.; Sen, S. Enhancement in Energy Storage and Piezoelectric Performance of Three Phase (PZT/MWCNT/PVDF) Composite. *Mater. Chem. Phys.* **2020,** *244*, 122639.
131. Ishaq, S.; Kanwal, F.; Atiq, S.; Moussa, M.; Azhar, U.; Imran, M.; Losic, D. Advancing Dielectric and Ferroelectric Properties of Piezoelectric Polymers by Combining Graphene and Ferroelectric Ceramic Additives for Energy Storage Applications. *Materials* **2018,** *11* (9), 1553.
132. Alam, M. M.; Ghosh, S. K.; Sarkar, D.; Sen, S.; Mandal, D. Improved Dielectric Constant and Breakdown Strength of γ-Phase Dominant Super Toughened Polyvinylidene Fluoride/$TiO_2$ Nanocomposite Film: An Excellent Material for Energy Storage Applications and Piezoelectric Throughput. *Nanotechnology* **2016,** *28* (1), 015503.
133. Chakhchaoui, N.; Jaouani, H.; Ennamiri, H.; Eddiai, A.; Hajjaji, A.; Meddad, M.; Van Langenhove, L.; Boughaleb, Y. Modeling and Analysis of the Effect of Substrate

on the Flexible Piezoelectric Films for Kinetic Energy Harvesting from Textiles. *J. Compos. Mater.* **2019,** *53* (24), 3349–3361.
134. Chakhchaoui, N.; Ennamiri, H.; Hajjaji, A.; Eddiai, A.; Meddad, M.; Boughaleb, Y. Theoretical Modeling of Piezoelectric Energy Harvesting in the System Using Technical Textile as a Support. *Polym. Adv. Technol.* **2017,** *28* (9), 1170–1178.
135. Karan, S. K.; Bera, R.; Paria, S.; Das, A. K.; Maiti, S.; Maitra, A.; Khatua, B. B. An Approach to Design Highly Durable Piezoelectric Nanogenerator Based on Self-Poled PVDF/AlO-rGO Flexible Nanocomposite with High Power Density and Energy Conversion Efficiency. *Adv. Energy Mater.* **2016,** *6* (20), 1601016.
136. Surmenev, R. A.; Orlova, T.; Chernozem, R. V.; Ivanova, A. A.; Bartasyte, A.; Mathur, S.; Surmeneva, M. A. Hybrid Lead-Free Polymer-Based Nanocomposites with Improved Piezoelectric Response for Biomedical Energy-Harvesting Applications: A Review. *Nano Energy* **2019,** *62*, 475–506.
137. Yu, Y.; Narita, F. Evaluation of Electromechanical Properties and Conversion Efficiency of Piezoelectric Nanocomposites with Carbon-Fiber-Reinforced Polymer Electrodes for Stress Sensing and Energy Harvesting. *Polymers* **2021,** *13* (18), 3184.
138. Bhattacharya, D.; Bayan, S.; Mitra, R. K.; Ray, S. K. 2D $WS_2$ Embedded PVDF Nanocomposites for Photosensitive Piezoelectric Nanogenerators with a Colossal Energy Conversion Efficiency of ~25.6%. *Nanoscale* **2021,** *13* (37), 15819–15829.
139. Jia, Y.; Wei, J.; Wang, K.; Cao, A.; Shu, Q.; Gui, X.; Zhu, Y.; Zhuang, D.; Zhang, G.; Ma, B.; Wang, L. Nanotube–Silicon Heterojunction Solar Cells. *Adv. Mater.* **2008,** *20* (23), 4594–4598.
140. Scarselli, M.; Scilletta, C.; Tombolini, F.; Castrucci, P.; De Crescenzi, M.; Diociaiuti, M.; Casciardi, S.; Gatto, E.; Venanzi, M. Photon Harvesting with Multi Wall Carbon Nanotubes. *Superlattices Microstruct.* **2009,** *46* (1–2), 340–346.
141. Hu, L.; Zheng, G.; Yao, J.; Liu, N.; Weil, B.; Eskilsson, M.; Karabulut, E.; Ruan, Z.; Fan, S.; Bloking, J. T.; McGehee, M. D. Transparent and Conductive Paper from Nanocellulose Fibers. *Energy Environ. Sci.* **2013,** *6* (2), 513–518.
142. Zhou, Y.; Fuentes-Hernandez, C.; Khan, T. M.; Liu, J. C.; Hsu, J.; Shim, J. W.; Dindar, A.; Youngblood, J. P.; Moon, R. J.; Kippelen, B. Recyclable Organic Solar Cells on Cellulose Nanocrystal Substrates. *Sci. Rep.* **2013,** *3* (1), 1536.
143. Fang, Z.; Zhu, H.; Yuan, Y.; Ha, D.; Zhu, S.; Preston, C.; Chen, Q.; Li, Y.; Han, X.; Lee, S.; Chen, G. Novel Nanostructured Paper with Ultrahigh Transparency and Ultrahigh Haze for Solar Cells. *Nano Lett.* **2014,** *14* (2), 765–773.
144. Kim, H. J.; Yim, E. C.; Kim, J. H.; Kim, S. J.; Park, J. Y.; Oh, I. K. Bacterial Nano-Cellulose Triboelectric Nanogenerator. *Nano Energy* **2017,** *33*, 130–137.
145. Liu, L.; Guo, Y.; Wang, Y.; Yang, X.; Wang, S.; Guo, H. Hollow NiO Nanotubes Synthesized by Bio-Templates as the High Performance Anode Materials of Lithium-Ion Batteries. *Electrochim. Acta.* **2013,** *114*, 42–4

# CHAPTER 4

# Sustainable Polymer Nanocomposites for Water Purification

JIJO ABRAHAM[1], SRILAKSHMI P. BHASKER[2], and SUSAN SAMUEL[2], NIMITHA K. C.[2,3], SALIYA C. S.[2], EUNICA K. SUNNY[2], and JIJI ABRAHAM[2]

[1]*St. Aloysius College, Elthuruth, Thrissur, Kerala, India*

[2]*Vimala College (Autonomous) Thrissur, Kerala, India*

[3]*Christ College (Autonomous), Irinjalakuda, Kerala, India*

## ABSTRACT

Rapid industrialization, urbanization, population growth, and climate change have increased the amount of wastewater being discharged, which contains diverse contaminants that impact living things. Several physical, chemical, and biological methods are adopted for water purification. Polymer nanocomposites have been considered an ideal and efficient material for water purification because of their amended processability, stability, surface area, tunable properties, high chemical reactivity, excellent mechanical strength, and cost-effectiveness. Polymer nanocomposites can be used to remove heavy metals, dyes, and organic and inorganic pollutants from water and also to desalinate water. This chapter summarizes the role of polymer nanocomposites in wastewater treatment.

Polymers and Functional Materials for a Cleaner Environment and Human Health: Sustainable Technologies and New Research. Ruby Varghese, Ann Rose Abraham, Shrikaant Kulkarni, & A. K. Haghi (Eds.)

## 4.1 A BRIEF INTRODUCTION TO POLYMER NANOCOMPOSITES

Polymer nanocomposite (PNCs) are very important materials where polymers are filled with nano inclusions. Polymers are formed by the combination of monomers and these have advantages such as light weight, flexibility, ease of production, and low cost. Nanofillers are defined as those materials with at least one dimension in the nano-size range. These nanofillers are classified into various categories based on the dimension, namely, 1-dimensional filler, 2-dimensional filler, 3-dimensional filler, etc. Nanofillers have excellent electric, dielectric, catalytic, magnetic, ionic, and chemical properties. The synergistic combination of polymers and nanofillers leads to improvement in the overall performance of PNCs. These include improved electronic, ionic, mechanical, and chemical properties that may be exploited for different applications.[1] Property enhancement mainly depends on the uniform dispersion of fillers in the polymer matrix and also the interfacial interaction between the filler and the polymer. There are several techniques to fabricate PNCs. Melt-mixing, mixing, in situ polymerization, solution casting, electro-spinning, etc., are the commonly used techniques to produce PNC.[2]

All living organisms require water. However, it is expected that less than 1% of the world's freshwater will be accessible, and pollution is one of the most serious environmental hazards. Presently, more than two billion individuals worldwide lack access to safe drinking water, as highlighted in the United Nations' Sustainable Development Goals report (SDG, 2022).[3] Additionally, the Intergovernmental Panel on Climate Change (IPCC) estimates that roughly half of the global population faces acute water scarcity issues during certain periods annually.[4] The increasing demand for clean water, driven by population increase, industrial growth, and the effects of climate change, is putting unprecedented pressure on world water supplies.[5] Water shortage is going to become a significant issue in the next decades, causing responses ranging from mass migration to other conflicts unless new methods of delivering clean water become accessible.[6] Fortunately, an increase in water treatment research in recent years gives promise for reducing the worldwide effect of polluted waterways. Many of these quality and supply issues may be solved using typical water treatment, decontamination, and desalination processes.[7]

## 4.2 METHODS OF WATER PURIFICATION

Water purification involves a diverse array of techniques that play a crucial role in guaranteeing accessible, safe, and clean water. Various physical processes, including filtration and sedimentation, utilize materials such as sand, gravel, and membranes to eliminate particles and solids. Chemical methods such as chlorination, ozonization, and advanced oxidation employ compounds to eliminate or break down pollutants such as pathogens and heavy metals. Biological treatment, commonly used in wastewater treatment plants and wetlands, harnesses microorganisms to degrade contaminants. Adsorption, utilizing materials such as activated carbon, captures contaminants through surface interactions. Ion exchange facilitates the removal of dissolved compounds such as calcium by exchanging ions. Distillation and reverse osmosis, through heat or pressure, separate pure water from impurities. Emerging technologies such as UV irradiation and electrochemical purification show promise in disinfection. These strategies, each with its own set of advantages and disadvantages, are frequently applied to address various water quality issues, ensuring the provision of safe water for communities worldwide.[8]

### *4.2.1 PHYSICAL ADSORPTION IN WATER PURIFICATION*

Physical adsorption is an important approach in water filtration that uses materials such as activated carbon and zeolites to attract and trap pollutants in water.[9] These materials, with their porous surfaces, provide enough space for pollutants to attach to, successfully eliminating organic chemicals, chlorine, volatile compounds, and some heavy metals.[10] Physical adsorption, which is widely employed in municipal water treatment facilities and housing filtration systems, efficiently cleanses water by utilizing the concept of pollutants sticking to a solid surface, assuring the supply of safer and cleaner drinking water. Physical adsorption is highly effective in removing organic compounds, chlorine, volatile organic compounds (VOCs), and certain heavy metals from water. Its versatility makes it applicable in various settings, including municipal water treatment plants, industrial processes, and household filtration systems.

Physical adsorption, however, has limitations. Over time, the adsorbent materials get saturated, reducing their effectiveness. Furthermore, certain pollutants may not be removed efficiently using this procedure, demanding the use of additional purification techniques.[11]

### 4.2.2 CHEMICAL DEGRADATION IN WATER PURIFICATION

Chemical degradation is an important aspect of water purification since it uses various chemicals to purify water. Contaminants are broken down or neutralized using methods such as chlorination, ozonation, and advanced oxidation.[12] Chlorination eliminates hazardous viruses and bacteria in water by using chlorine. Ozonation employs ozone to convert toxic substances in water into less harmful ones. Special molecules are used in advanced oxidation to break down complex harmful substances in water. These procedures are widely employed in water treatment plants to ensure that people have clean drinking water. However, employing chemicals such as chlorine can occasionally result in water that is hazardous to one's health. Furthermore, several of these processes need a significant amount of energy or careful management of chemicals. Despite these obstacles, chemical approaches remain essential.

### 4.2.3 BIOLOGICAL TREATMENT IN WATER PURIFICATION

Biological treatment relies on the inherent capabilities of microorganisms to naturally degrade and eliminate water pollutants. This eco-friendly method harnesses the metabolic processes of various living organisms, including fungi, bacteria, and algae, to break down organic materials and remove pollutants. Processes such as biofiltration, artificial wetlands, and activated sludge play pivotal roles in facilitating the microbial breakdown of organic contaminants. Activated sludge systems, in particular, rely on microorganisms that feed on organic materials, contributing significantly to the breakdown of pollutants. This method effectively removes organic molecules, nutrients, and certain heavy metals from water. Its application extends beyond conventional wastewater treatment facilities, finding utility in decentralized systems and natural settings for water filtration. However, the efficiency of biological treatment is subject to various influencing factors, such as temperature, pH levels, and the specific types of contaminants present. The success of the treatment depends on the intricate balance of these variables. Despite its environmental benefits, it's noteworthy that biological treatment may need extended treatment durations when compared to alternative approaches.[13]

## 4.3 APPLICATION OF POLYMER NANOCOMPOSITES FOR WATER PURIFICATIONS

Rapid industrialization, urbanization, population growth, and climate change have increased the amount of wastewater being discharged, which contains diverse contaminants that impact living things. Traditional materials and treatment methods are ineffective for treating complex and intricately contaminated streams that contain numerous pollutants. Recently polymer nanocomposites have been considered as an ideal and efficient material for water purification because of their amended processability, stability, surface area, tunable properties, high chemical reactivity, excellent mechanical strength, and cost-effectiveness.[14] PNCs can be used to remove heavy metals, dyes, and organic and inorganic pollutants from water and also to desalinate water. This chapter summarizes the role of PNCs in wastewater treatment.

### *4.3.1 REMOVAL OF HEAVY METAL IONS*

The term "heavy metals" refers to elements with an atomic mass between 63.5 and 200.6 and a specific gravity greater than 5.0 (e.g., As, Hg, Cr, Pb, Ni, and Co).[15] Since they are persistent pollutants and nonbiodegradable, drinking water contaminated with heavy metal ions (HMIs) causes toxicity and major health problems in living organisms. HMIs can be removed from contaminated water using several methods, including precipitation, ion exchange, adsorption, membrane filtration, electro-deposition, reverse osmosis, and solvent extraction.[16] However, a great number of research studies have been reported that address major shortcomings of existing methods such as inadequate or partial removal of HMIs, limited selectivity, secondary waste production, and high operational cost. Adsorption, which uses a variety of adsorbent materials including agricultural residue, inorganic, polymeric, and composites, stands out as a better approach due to its ease of use, environmentally benign nature, and low operational cost.[17] On the other hand, the recovery of adsorbents from treated water is a major concern. To overcome this issue, PNCs are currently being investigated for application as adsorbents or as adsorptive ultrafiltration membranes.[18] They are characterized by many desirable properties such as ease of manufacture, cost-effectiveness, environmental stability, and binding sites with substantial surface area and pore volume. A categorized discussion on different types of PNCs developed for the removal of HMIs is given in the following sections.

#### *4.3.1.1 CLAY-BASED POLYMER NANOCOMPOSITES*

Clay minerals are hydrous aluminosilicate layers composed mainly of tetrahedral $(SiO_4)^{4-}$ and octahedral $[Al_2O_3(OH)_3]^{3-}$ sheets (e.g., montmorillonite, kaolinite, and zeolite).[19] Because of their low cost, environment-friendly nature, abundance, high surface area, high cation exchange capacity, applicability in a wide pH range, and high chemical stability, clay minerals are explored well as adsorbents for HMIs.[20] When added in small quantities to a polymer matrix, they can cause a significant improvement in the properties of polymers. The adsorption of HMIs by clay-polymer composites is primarily driven by electrostatic attraction and ion exchange processes operating between the active sites of the composite and the HMIs. Montmorillonite, a layered clay mineral, is a good choice for HMIs' removal. The surface of montmorillonite possesses a negative charge, similar to that of a large anion, which is neutralized by the adsorption of a large number of cations. Thus, montmorillonite can be used for HMI adsorption via ion exchange replacing interlayer cations present in the structure. In a recent study, montmorillonite grafted with sodium polyacrylate was used for the adsorption of Cd ions from water.[21] A step-by-step procedure employed for the preparation of polymer-grafted montmorillonite is given schematically in Figure 4.1. As shown in Figure 4.1, the preparation of composites involves the grafting of montmorillonite with a silane coupling agent at first which is followed by extended polymerization in the second step. Here, initial grafting is performed by the dehydration condensation reaction between Si/Al-OH on the clay surface and the silicon hydroxyl group produced by the hydrolysis of the silane coupling agent. The polycarboxylate groups on the modified montmorillonite greatly increased

Cd ion adsorption ability, yielding a maximum adsorption capacity of 63.49 mg/g. In another study, a magnetically separable polyethylenimine (PEI)-functionalized montmorillonite clay (MMT-$Fe_3O_4$-PEI) adsorbent was developed for Cr(VI) removal from aqueous solution.[22] The electrostatic attractions between protonated amino groups on the polymer matrix with Cr(VI) anions followed by partial reduction of Cr(VI) to Cr(III) resulted in the removal of Cr ions. A maximum adsorption capacity of 62.89 mg $g^{-1}$ was reported for this reusable magnetic clay-polymer composite.

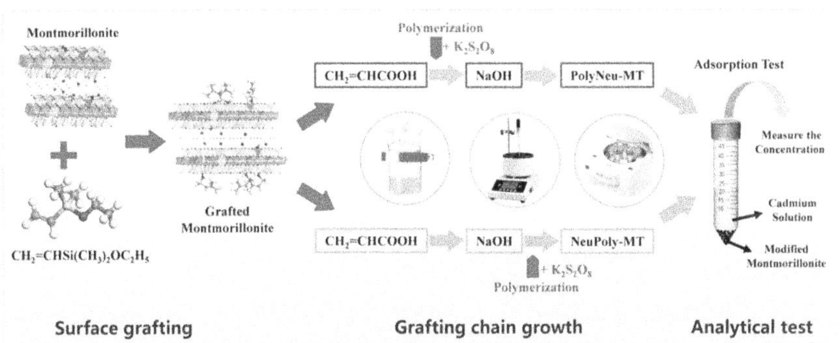


**FIGURE 4.1** Schematic representation of steps involved in the preparation of polyacrylate grafted montmorillonite for Cd ion removal.

*Source:* Reproduced with permission from Ref. [21]. Copyright © 2023. Elsevier.

Kaolinite is another low-cost and abundant clay mineral with a high surface area and high cation exchange capacity. For instance, kaolinite-based cellulose/$Co_3O_4$ composite was developed as a green adsorbent for Pb(II) and Cd(II) decontamination from an aqueous solution.[23] The plausible mechanism of Pb(II) and Cd(II) removal by the composite is illustrated in Figure 4.2. As shown in Figure 4.2, the decontamination process is believed to entail the combined effect of electrostatic, ion exchange, and chelation mechanisms. Here, various functional groups present in of the composite functions as the active sites for HMIs' sorption. In another work, carboxymethyl chitosan-kaolinite composite hydrogel was used for Cu(II) trapping.[24] The addition of kaolinite improved the mechanical properties of the hydrogel which in turn facilitated the dispersion and separation of adsorbents.

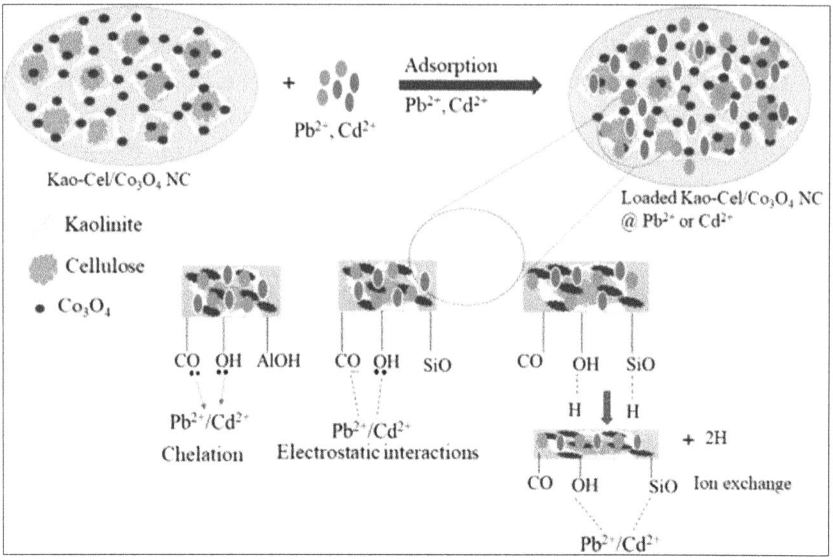


**FIGURE 4.2** Illustration of plausible mechanism responsible for the removal of Pb(II) and Cd(II) with kaolinite-cellulose/$Co_3O_4$ composite.

*Source:* Reproduced with permission from Ref. [23]. Copyright © 2022 The Author(s). Published by Elsevier B.V. on behalf of King Saud University.

### *4.3.1.2 METAL AND METAL OXIDE-BASED POLYMER NANOCOMPOSITES*

Metal-based systems in conjunction with conducting polymers (e.g., polyaniline (PANI), polypyrrole (PPY), and polythiophene) are extensively studied for the removal of HMIs. Conducting polymers have electron-donating groups in their structure with which they can chelate HMIs. In addition to that, a positive charge can be introduced on the amine nitrogen atoms of the polymer via doping which in turn would help in the removal of negatively charged pollutants via electrostatic interaction.[25] PANI can chelate metal ions with the help of imine and amine groups and it can adsorb metal ions via electrostatic or hydrogen bonding. $NiFe_2O_4$–PANI composite,[26] core-shell $ThO_2$-PANI composite,[27] and PANI/Fe(0) composite[28] are examples of adsorbents developed using PANI for the removal of various HMIs. Likewise, other conducting polymer-based systems such as PPy/$TiO_2$ composite[29] and poly-thiophene/$TiO_2$ composite[30] are also known for Cu(II) removal from aqueous solution.

Nowadays, iron oxide ($Fe_3O_4$) is studied extensively as one of the components of composite structure because of its nontoxic nature, high affinity toward contaminants, and possible magnetic separation. For instance, a linear biopolymer, inulin extracted from Jerusalem artichoke tubers with a low-cost method was employed for HMIs removal in the form of a magnetic composite with $Fe_3O_4$.[31] Using 0.01g adsorbent at pH 5.5, a maximum adsorption capacity of 152.5, 167.7, and 198 mg/g was reported for $Co^{2+}$, $Cu^{2+}$, and $Hg^{2+}$, respectively. Similarly, PPY- $Fe_3O_4$ seaweed nanocomposite was proposed for Pd(II) removal from water.[32] Here, seaweed played a dual role of encapsulating $Fe_3O_4$ preventing self-aggregation of $Fe_3O_4$, and functioned as a platform for the incorporation of PPy. Other examples of iron oxide-based systems are chitosan/$Fe_3O_4$ composites,[33] PPY-modified $Fe_3O_4$/reduced graphene oxide (rGO) composites,[34] $Fe_3O_4$@polydopamine@polyamidoamine composites,[35] and polyvinylpyrrolidone/$Fe_3O_4$ composites.[36]

Another factor to consider while developing adsorbents is the selective adsorption of HMIs in the presence of interfering cations such as $Na^+$, $K^+$, $Mg^{2+}$, and $Ca^{2+}$, which are typically found in water. Saravaia et al. have developed polysulfone-supported magnesium-doped lithium manganese oxide nanosorbent beads for the selective removal of HMIs in the presence of other interfering cations.[37] Unlike unsupported nanosorbent, the selective removal was limited to $Pb^{2+}$, $Cu^{2+}$, and $Zn^{2+}$ in the presence of polymer. The suggested explanation for this restricted performance of PNC was the possible competition and polymer obstruction on the sites responsible for other HMIs' adsorption.

#### 4.3.1.3 CARBON-BASED POLYMER NANOCOMPOSITES

The insertion of carbon-based nanomaterials into the polymer matrix can enhance the performance of polymer membranes by improving their properties, particularly their hydrophilicity, fouling resistance, and permeability. Among various allotropes of C, graphene oxide (GO) has received increased attention as a multivalent metal ion adsorbent due to its large surface area, hydrophilicity, and high negative charge density.[38] For instance, a mixed matrix membrane of high porosity and thermal stability was developed using carboxylated graphene oxide (c-GO) microsheets and polyetherimide.[39] The introduction of carboxyl group on GO microsheets improved the hydrophilicity of the adsorption membrane significantly. A liquid phase retention approach was used in this study to remove HMIs. The chelating polymer forms complexes with the target metal ions that can be efficiently held by ultrafiltration membranes. For membranes loaded with 0.3 wt % c-GO, Ni and Cd removal was estimated to be 95.5% and 92.4%, respectively. On the other hand, adsorbent beads for the removal of Pb (II), Hg (II), and Cd (II) were developed with GO embedded in cross-linked alginate matrix (GOCA).[40] To increase the adsorption capacity further, the developed beads were functionalized with PEI which also led to partial reduction of GO. The synthesis and functionalization of GOCA beads are shown schematically in Figure 4.3. Maximum adsorption capacity of 602 mg/g for Pb, 374 mg/g for Hg, and 181 mg/g for Cd was demonstrated by functionalized GOCA beads. The high adsorption capacity of beads is attributed to the availability of oxygen functionalities in GO, amine functional groups on PEI, and carboxylate groups of alginate for complexation with metal ions. Furthermore, encapsulating GO onto calcium alginate beads eliminates particle leaching and aggregation.

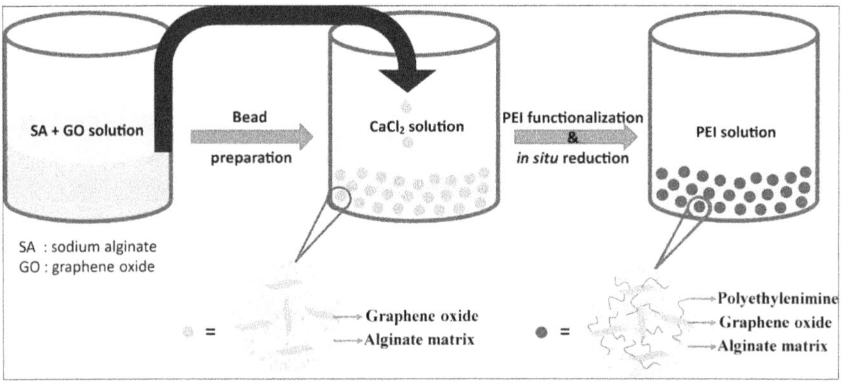


**FIGURE 4.3** A schematic representation of preparation of PEI-functionalized GOCA beads.
*Source:* Reproduced with permission from Ref. [40]. Copyright © 2019. Elsevier.

In another study, PEI modified with multiwalled carbon nanotube (MWCNT) was used as an adsorbent for Cr(VI) removal from contaminated water.[41] The use of unmodified carbon nanotubes is restricted due to their low dispersibility and lack of surface functional groups, which can be resolved via functionalization with PEI. Chen et al. reported a PEI-modified MWCNTs@$Fe_3O_4$ adsorbent for the simultaneous removal of methyl orange and Cr(VI) from aqueous solutions.[42] The impressive performance of the composite has been attributed to the high surface area of the MWCNTs, the abundance of protonated N-rich groups of PEI, and the good dispersibility of nanocomposites.

### 4.3.2 REMOVAL OF DYES BY POLYMER NANOCOMPOSITES

Nonscientific treatment of wastewater may lead to several kinds of effluents that flow to natural water bodies. The effluents are consisting of HMIs, dyes such as methylene blue (MB) and rhodamine B, and distillate microcystins. Several chemical, physical, and biological methods are utilized for the removal of these contaminants from wastewater. The presence of dyestuffs in water streams adversely affects the ecological system.[43] Approximately, 1,000,000 tons of dyes are consumed globally because of activities such as tanning, food processing, and manufacturing of cosmetics, medicines, and textiles. Dyes that are used in current industries consist of complex aromatic ring systems and the color in them is due to the presence of chromophores and auxochromes. The functional groups present in dyes can absorb UV and visible light in the range of 380–700 nm and this will impart color to them. Several natural dyes are currently used. Synthetic dyes consisting of functional groups such as azo which become carcinogenic due to the emission of amines and benzidine. The nonbiodegradable nature of these dyes makes it a threat to the development of a sustainable environment. Therefore, the removal of dyes from the industrial effluents before discharging them to the environment is of paramount importance.[44]

Among the recent advancements in this field, one of the efficient ways to remove dyes from wastewater is adsorption and the other method is photodegradation. The adsorption method is comparatively widely used because of its low cost and simplicity. In this method, dyes are chemically or physically adsorbed onto the surface of suitable adsorbents. An adsorbent should possess characteristics such as high surface area, high adsorption capacity, and high selectivity and should be economical, compatible, and easily regenerative, too. Adsorption is not sufficient to detoxicate the dyestuff. It just removes the dyes from the wastewater. But photodegradation provides an efficient way to completely decompose the pollutant using a photocatalyst under sunlight or UV light. The efficiency of photodegradation depends on several factors such as concentration of pollutant and catalyst, pH, and physical parameters such as temperature, intensity, wavelength of light used, etc. Several materials can act as highly efficient adsorbents and photocatalysts. PNCs are potential candidates as adsorbents and photocatalysts since they possess high surface area and porosity. Conducting PNCs of polymers such as polyaniline, polypyrrole, polythiophene, and so on, is widely utilized as adsorbents and photocatalysts to remove organic dyes from wastewater. Clay-based nanoparticles, carbon-based nanomaterials, and metal and metal oxide nanoparticles can be used to make PNCs for the applications.

#### *4.3.2.1 CLAY-BASED POLYMER NANOCOMPOSITES*

Clay minerals are consisting of phyllosilicates that are arranged in layers. Each layer is composed of octahedral $[AlO_3(OH)_3]^{6-}$ and tetrahedral $SiO_4^{4-}$. Clay minerals can be used to make PNCs and can be efficiently used to remove both organic and inorganic pollutants. Generally, polymers such as polystyrene, polyurethanes, polypropylene, polyvinyl chloride, and chitosan are used to make composites with clay minerals. Chitosan/bentonite-based clay PNC is used as an adsorbent in removing a very harmful azodye tetrazine. Studies have shown that the rate of adsorption is maximum at low pH because, at low pH, a positive charge is developed on the surface of clay PNC (CPN), and this facilitates the electrostatic attractions between dye and nanocomposites. Chitosan-based clay magnetic bead nanocomposites are effectively used for the fast removal of MB. Very reactive organic dyes such as yellow K-4G and dispersed yellow-brown S-2RFL are removed by using a clay-based nanocomposite of poly(epicholorohydrin-dimethylamine) as adsorbent.[45] Cationic dyes such as malachite green (MG), MB, and crystal violet (CV) from water are removed using a humic acid immobilized-amine modified bentonite–polyacrylamide (HA-Am-Bent–PAA) nanocomposite. Clay-based PNCs can also be used to remove dye mixtures. For example, bentonite modified with poly(epichlorohydrin dimethylamine) is used to remove anionic dyes such as direct fast scarlet, eosin Y, and reactive violet K- 3R from single, binary, and ternary dye mixtures.[46] Removal of aqueous secondary diazo dye, Congo red, using CPN from water was also reported. The mechanistic studies show that the adsorption of Congo red on CPN follows chemisorption. Another important finding is the removal of very commonly used dye, that is, rhodamine-6G using chitosan (CTS)-clay nanocomposite. The following process shows a 446.42 mg/g of dye adsorption potential which is quite high. The mechanism follows a pseudo-second-order kinetics.[47] A PNC of clay-based minerals and cetyltrimethyl ammonium bromide-modified bentonite intercalated with polyacrylamide showed extremely good adsorption capacity towards phenol and nitrobenzene. Removal of dyes such as reactive blue-21 and reactive red-141 using a chitosan-modified nano-organo clay was also reported by Vanaamudan and Sudhakar. Kinetic studies ensure approximately 60–62% of dye removal by this PNC within a timespan of only 50 minutes.[48] Starch-montmorillonite/polyaniline (St-MMT/PANI) nanocomposite which is an eco-friendly clay-based polymer composite is also reported as an efficient adsorbent for a reactive dye.[49]

#### *4.3.2.2 METAL AND METAL OXIDE-BASED POLYMER NANOCOMPOSITES*

Several metal and metal oxide-based PNCs can be effectively used for the removal of dyes from wastewater effluents. A novel iron oxide/polyaniline PNC (iron oxide-PANI-PNC) is prepared by simply impregnating onto polyaniline and can be used as a better adsorbent to remove eriochrome black T dye from the water matrix. The mechanism of adsorption is found to be sono-assisted in this case.[50]

A polypyrrole-modified $TiO_2$ is used to adsorb dyes such as acid red G and MB from the aqueous matrix. Vast research has been carried out to investigate the adsorption characteristics of various dyes. Adsorption of MB by PNCs of metal oxides such as PPy/$SiO_2$, PPy/$Al_2O_3$, and PPy/$Fe_3O_4$ is thoroughly studied.[51] The surface area of metal and metal oxide-based PNCs plays a vital role in their adsorption efficiency. Studies show the degradation of MB dye using a zinc oxide-cellulose nanocomposites. The prepared material shows exceptionally high degradation efficiency than pristine ZnO. Microfibrillated cellulose (MFC) composites of $TiO_2$ shows good degradation rate toward methyl orange dye. Cellulose polymer is the most abundant natural polymer and therefore scope for the development of a stable, low cost and recyclable, and biocompatible metal-based PNC is endlessly high.[52] Figure 4.4 represents UV-visible spectra of photocatalytic degradation of MB dye using ZnO and ZnO-cellulose nanocomposites at different time intervals. The extend of degradation is more in cellulose-ZnO composites and is evident from the results.[53]

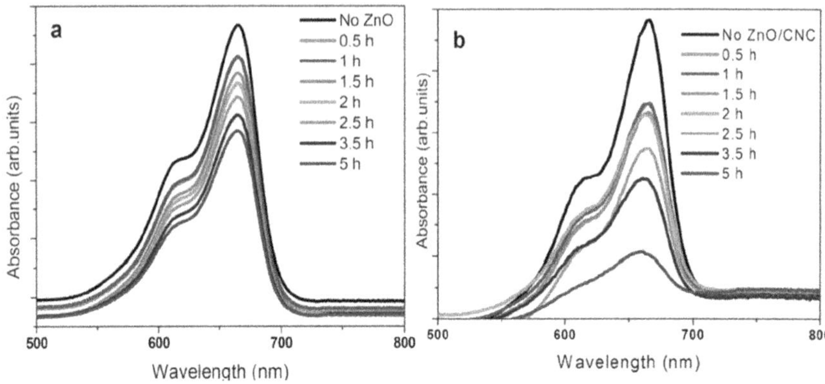


**FIGURE 4.4** Photocatalytic degradation of methylene blue dye using (a) ZnO NPs and (b) ZnO-cellulose nanocomposites.

*Source:* Reproduced with permission from Ref. [53]. Copyright © 2017. Elsevier.

### 4.3.2.3 CARBON-BASED POLYMER NANOCOMPOSITES

Several carbon-based PNCs are widely used to remove the dyes from the aqueous systems and are widely used for wastewater treatment. A PNC of GO and polyaniline (PANI) was used for the adsorptive removal of brilliant green (BG) from an aqueous solution.[54] A branch such as carboxylated MWCNTs with chitosan polymer membrane effectively shows rejection of dyes such as MB and methyl orange. Another study shows the removal of direct red 16 (DR16) by PES/ZnO MWCNT nanocomposite membrane.[55] Ghaemi et al. showed the removal of acidic orange 7 by a PNC of MWCNTs with poly(sodium-4-styrenesulfonate) (PSS).[56] Nanofiltration using a suitable filter membrane such as CNT/polymer mixed matrix can be used for the dye removal. Another example is the incorporation of GO into poly(m-phenylene isophthalamide) (PMIA). The prepared material shows good removal of dyes such as cationic red-gtl (RG), acid red (AR), and reactive yellow (RY) through nanofiltration. A copolymer of aniline and indole P(ANI-co-IN) is prepared with GO and carbon nanotubes (CNT) were synthesized and found high efficiency for the removal of cationic dyes such as MB and MG from water bodies.[57] A polymer composite of carbon black and polyvinyl alcohol (PVA) is utilized well for the removal of aniline yellow by adsorption.[58] Polymers such as poly-(4,4′-oxybisbenzenamine) (POBBA) and poly-(4,4′-oxybisbenzenamine-pyrrole) (COP) with carbon dots act as free radical initiators for the removal of different dyes such as congo red, crystal violet, MB, and rhodamine B from aqueous solution.[59] Figure 4.5 represents the adsorption mechanism of BG on a polymer composite of GO-PANI.

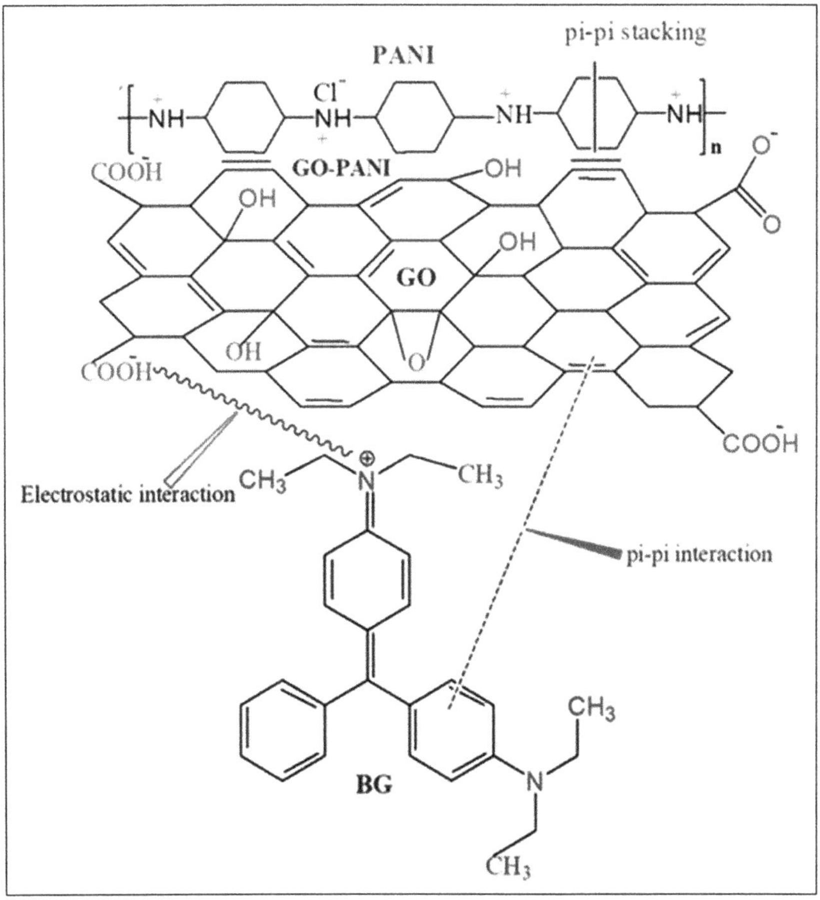


**FIGURE 4.5** Brilliant green (BG) adsorption mechanism on GO-PANI.

## 4.3.3 REMOVAL OF OTHER POLLUTANTS

### 4.3.3.1 CLAY-BASED POLYMER NANOCOMPOSITES

It is worth mentioning that, in general, clay-based materials are commonly applied as effluent pre-treatment and pollutant removal. CPNs can adsorb by either inorganic and organic micropollutants. CPNs are widely used in effective water treatment systems. Additionally, they provide a great price advantage compared to other adsorbents, namely, zeolites, activated carbon, and so on. Figure 4.6 shows the wastewater treatment using CPNs.[60]

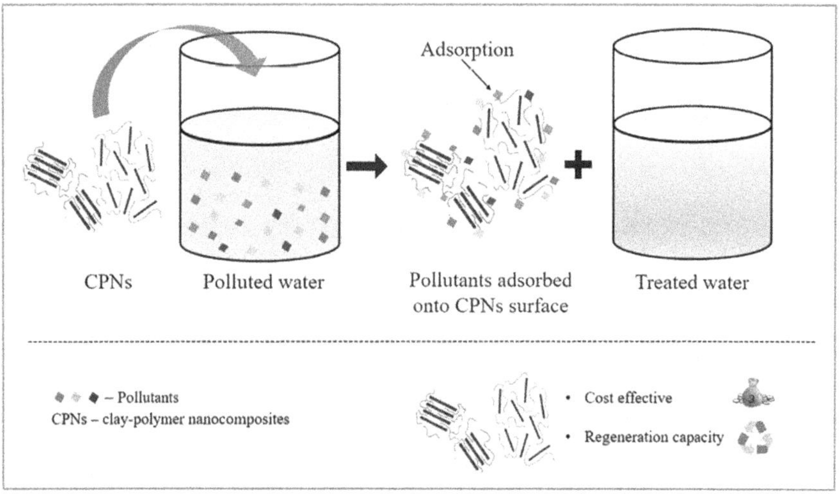


**FIGURE 4.6** Removal of pollutants from polluted water using clay-based polymer nanocomposites.

*Source:* Reproduced with permission from Ref. [19]. Copyright © 2022, Elsevier.

CPNs such as montmorillonite-polydimethyloxane-chlorhexidine acetate and clay-polydimethyloxane-chitosan-silver exhibited the ability to slow down the growth of different bacteria (*E. coli*, *Pseudomonas aeruginosa*, and *Staphylococcus aureus*).[61–63] A composite, synthesized from acid-activated bentonite and natural palygorskite via polymerization reaction between N,N'-methylenebisacrylamide and acrylic acid, was used for the adsorption of ammonia in wastewater treatment systems.[64] Phenolic materials such as 4-chlorophenol and phenols were removed using alginate polymer support impregnated by clay modified with a surfactant hexadecyl trimethyl ammonium. This adsorbent showed a high removal capacity for 4-chlorophenol and phenols.[65]

Several other phenolic materials were eliminated by a nanocomposite prepared using cetyltrimethylammonium poly(diallyldimethylammonium) to modify montmorillonite clay. Although the surfactant was not properly injected inside the clay montmorillonite, the adsorption capacity for phenols was drastically enhanced.[66] Unuabonah et al. showed that a CPN from kaolinite and polyvinyl alcohol (Kaol–PVA) can efficiently be used in the elimination of lead and cadmium from aqueous solution.[667] Ganigar et al. used montmorillonite–poly-4-vinylpyridine-co-styrene (Mt–PVPcoS) and montmorillonite–polydiallydiamethylammonium (Mt–PDADMAC) nanocomposite for the removal of trichlorophenol and trinitrophenol from industrial waste water.[68] Wang et al. also found that novel CPN from cetyl-trimethyl ammonium bromide (CTAB) and modified rectorite had great antimicrobial capacity in both weakly acidic and basic aqueous medium against a wide range of Gram-positive and Gram-negative microorganisms.[69]

Clay-polymethoxyethylacrylamide (PMEA) nanocomposites are successfully used to remove Pb(II) from an aqueous solution with a very high adsorption capacity.[70] Zhu et al. also prepared another clay-based nano composite from polymer/surfactant-modified montmorillonite (montmorillonite-cetyltrimethylammonium-poly(diallyldimethylammonium) Mt-Ct-PDADMA) and was used to remove phenol from water.[71] Wang et al. showed that modifying bentonite with cetyltrimethyl ammonium bromide and intercalating with polyacrylamide resulted in CPN whose efficiency for the adsorption of phenol and nitrobenzene from industrial wastewater was significantly enhanced by 26% and 28%, respectively.[72] Anirudhan et al. showed that humic acid immobilized-amine-modified bentonite–poly-acrylamide (HA-AmBent–PAA) nanocomposite can be used to effectively remove Cu, Zn, and Co from aqueous solutions.[73] Combinations of clay mineral sepiolite and organic quaternary ammonium ions along with linear cationic polymer poly-DADMAC as the bridging ribbon are used effectively for the removal of organic contaminants from olive mill wastewater.[74]

#### *4.3.3.2 METAL AND METAL OXIDE-BASED POLYMER NANOCOMPOSITES*

There are different types of nanomaterials such as Ag, Ti, and Zn capable of removing various pollutants in water since; when the particle size decreases to nanoscale, they become more efficient owing to the large surface area and high reactivity. Several environmentally benign inorganic components, namely, metal oxides and hydroxides such as Fe,[75] Mn,[76] Zr, Ti, and Sn[77] have been used as efficient materials for enhanced elimination of pollutants in wastewater treatment applications. A new nanocomposite prepared by the association of nanosized hydrated ferric oxide into cross-linked polystyrene has been used to remove phosphate and p-nitrophenol from water.[78] Ag/MWNTs coated on a polyacrylonitrile (PAN) hollow fiber membrane, Ag/MWNTs/PAN, has been developed to enhance the antimicrobial activities and antifouling properties of the membrane against *Escherichia coli*.[79] Figure 4.7 depicts the immobilization of AgNP/CNTs coated on the surface of a polyacrylonitrile (PAN) hollow fiber membrane for water purification.

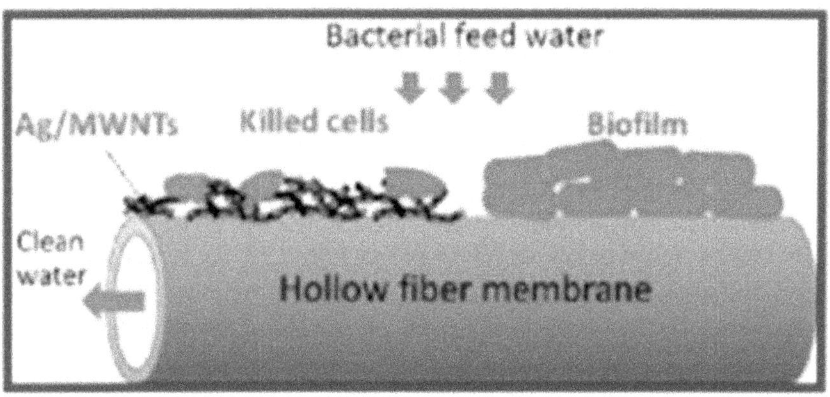


**FIGURE 4.7** Schematic representation of representative disinfection method.

*Source:* Reproduced with permission from Ref. [79]. Copyright © 2011, American Chemical Society.

Silver nanoparticles coated on polyurethane foams can be used as a drinking water filter. At an optimal flow rate, the output count of *E. coli* was nil when the input water had a high bacterial load.[80] A nanocomposite of polypyrrole/carbon nanotubes/Ag was prepared through the in situ oxidative polymerization of pyrrole with $AgNO_3$ containing SWCNT and achieved higher *E. coli* removal percentage.[81] There are different types of nanomaterials such as Ag, titanium, and zinc capable of disinfecting several waterborne disease-causing microbes. $TiO_2$ photocatalysts, metallic, and its oxide nanoparticles are among the most useful nanomaterials with antimicrobial properties owing to their charge capacity.[80,82] A thin sheet of nanocomposite made from titanium dioxide ($TiO_2$)-halloysite nanotubes (HNTs)/polyvinylidene fluoride (PVDF) membrane are utilized in the photocatalytic membrane reactor for the removal and degradation of different hydrocarbons from engine oil and fuel tank of ships, which, when discharged to the sea without proper treatment, contributes to major pollution in seawater and seriously affects marine ecosystems.[83] Munnawar et al. constructed the chitosan/zinc oxide nanocomposite and optimized it into a polyethersulfone matrix to fabricate antifouling polyethersulfone membranes and it showed excellent water permeability and prevented microbial fouling.[84] Figure 4.8 shows effective water disinfection strategy used by incorporating Zn oxide nanoparticles.

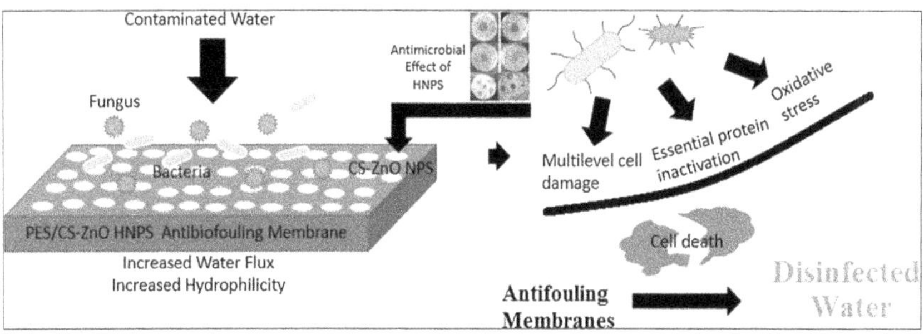


**FIGURE 4.8** Chitosan-zinc oxide hybrid nanoparticles-CS-ZnO HNPS with polyethersulfone matrix investigated for water disinfection applications.

*Source:* Reproduced with permission from Ref. [84]. Copyright © 2017. Elsevier.

Pesticides including insecticides, fungicides, rodenticides, and plant growth regulators which can reach surface water through runoff from treated plants and soil cause serious water pollution and removal of one of the pesticides permethrin has been carried out efficiently by chitosan zinc oxide nanocomposite.[85] Al-Naamani et al. made use of a chitosan/ZnO nanocomposite coating to prevent marine biofouling. The nanocomposite used showcased antibacterial activity against the marine bacterium *Pseudoalteromonas nigrifaciens* and anti-diatom activity against *Navicula* sp.[86] Removal of numerous radioactive nuclides discharged from active nuclear power plants is essential as they induce serious health problems such as cancer, anemia, osteoporosis, cataracts, kidney disease, and so on. Nano iron-Fe° and its composite with GO have effectively been used for the elimination of uranium from water.[87,88] Hydrated ferric oxides are very much effective in the field of wastewater treatment as they can bind anionic ligands such as phosphate and arsenate selectively through inner-sphere complex formation.[89,90] In another research analysis, polycaprolactone nanocomposite membrane with zinc oxide NPs was fabricated and it exhibited enhanced antimicrobial activity against *S. aureus* and *E. coli*.[91]

Sarkandi et al. fabricated a nanocomposite hydrogel comprising of cellulose and silver nanoparticles and exhibited excellent antibacterial with a complete reduction in bacterial percentage.[92] Several low-cost potable water microfilters were prepared by incorporating Ag nanoparticles and used in remote areas of developing countries.[93]

#### *4.3.3.3 CARBON-BASED POLYMER NANOCOMPOSITES*

Carbon nanomaterials, namely, MWCNT and single-walled carbon nanotubes (SWCNTs), activated carbon, graphene, GO-based nanomaterials, and carbon and graphene quantum dots (GQDs) have been used effectively as adsorbents for wastewater treatment and purification for the removal of pharmaceuticals, endocrine disrupting chemical materials, common industrial wastes, pesticides personal care products, detergents, and other emerging pollutants.[94] Figure 4.9 depicts the wastewater treatment using carbon-based nanomaterials.

CNT-based composites show remarkable wastewater remediation capabilities for numerous organic/inorganic and biological water contaminants.[95] SWCNTs and MWCNTs are now readily available for incorporation into several matrices to fabricate CNT/polymer composites. SWCNTs show higher removal capability for natural organic matter.[96] CNTs are considered as very good adsorbents and have attracted considerable attention for eliminating organic matters such as herbicides[97] and aromatic hydrocarbons.[98] The incorporation of MWCNTs into the aromatic polyamide membranes forms a nanocomposite structure with improved mechanical enhanced their ability to reject key contaminants.[99] Graphitic carbon nitride (g-C3N4)/carbon dots (CDs)-based nanocomposites recently obtained great research interest due to their favorable application in the removal of a wide variety of contaminants.[100] Zhang et al. synthesized $Fe_2O_3$/carbon quantum dots

(CQDs) nanocomposites as efficient photocatalysts for the degradation of poisonous gases namely benzene or methanol in visible light.[101]

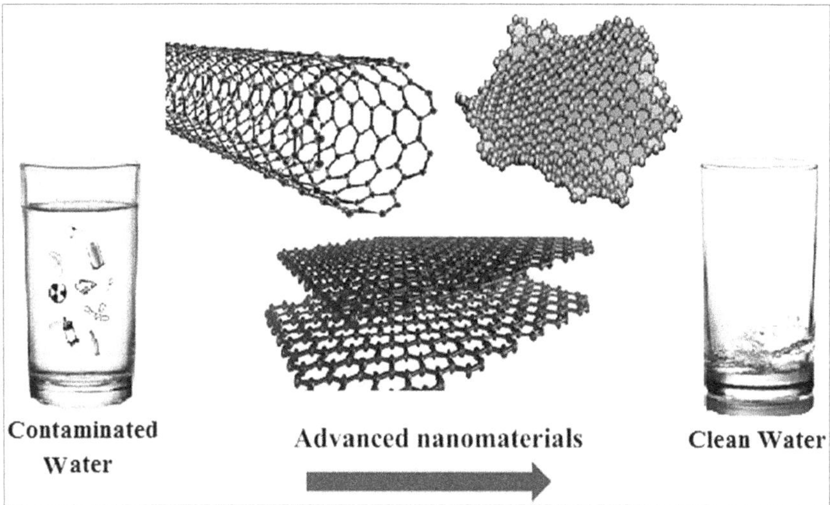


**FIGURE 4.9** Wastewater treatment using carbon-based nanomaterials.

*Source:* Reproduced with permission from Ref. [94]. Copyright © 2021. Elsevier.

CNTs, and carbon allotropes with a cylindrical nanostructure, can be used for eliminating the pharmaceutically derived contaminants.[102,103] Nanocomposite of SWCNT in polylactic-co-glycolic acid and poly-lysine polymers deactivates bacteria in a higher removal percentage in aqueous solution.[104,105] Nanocomposites of poly-N vinyl carbazole (PVK) with SWCNT (PVK-SWCNT) and graphene oxide (PVKGO) cause higher cell inactivation of both Gram-positive and Gram-negative bacteria in aqueous solution.[106,107] N-doped ZnO/CQDs exhibited super removal of a vast range of pollutants from industrial wastewater.[108] Zhang et al. fabricated gold NPs/GQDs nanohybrids that exhibited a synergetic and switchable peroxidase-like ability for $H_2O_2$ decomposition.[109] Chen et al. showed nanocomplex-based photocatalyst $WO_3$/CQDs have been revealed with substantial proficiency toward the photodegradation of a variety of organic contaminants.[110] Elanchezhiyan et al. exhibited GO decorated with lanthanum-substituted manganese ferrites (LMF) have been constructed for the removal of perfluorooctanoic acid from aqueous environments.[111]

### 4.3.4 REMOVAL OF OIL BY POLYMER NANOCOMPOSITES

Environmental pollution and other environmental problems are due to the rapid growth of industries. Energy resources such as petroleum and petroleum hydrocarbons are major pollutants of the environment.[112] One of the most pressing environmental issues facing the world today is oil pollution, which is brought on by the petrochemical, textile, and food sectors, as well as by the numerous mishaps that result in oil spills during offshore oil production or maritime shipping.[113] The oil and oil products contamination mainly affects water resources and may cause severe harm. Hence attention has been paid to the development of alternative methods for the removal of these pollutants. Two methods are mainly used for this purpose (i) using hydrophobic materials directly as oil adsorbent for oil/water separation and (ii) using membrane fibers coated with hydrophobic materials for the removal of oil from water.[114] Filtration by the polymeric membrane is an effective method used in wastewater treatment. But it has some limited applications. The filtration process can be enhanced by using nanocomposite membranes, which are formed by adding nanofillers to polymeric membrane matrices. The inclusion of nanomaterials in the polymer matrix has been increasing the chemical, mechanical, and thermal properties.[115]

#### 4.3.4.1 REMOVAL OF OIL BY CLAY-BASED POLYMER NANOCOMPOSITES

Clays are used as adsorbents for many micropollutants. Hence, clays can be easily manipulated and produced in a flexible structure; polymers were used in a composite with clays. The iron oxide/bentonite nanocomposites ($Fe_3O_4$) ($Fe_3O_4$/Bent NC), prepared by the ultrasound-assisted co-precipitation method, are a suitable nano-adsorbent for the removal of oil from a water–oil emulsion. Figure 4.10 illustrates the mechanism of oil removal using $Fe_3O_4$/Bent NC.[116]

The incorporation of cloisite 20A nano clay into polyurethane foam structure enhances the removal of oil contaminants from water. The characterization studies show that the oil sorption capacity has increased in low initial oil concentrations, while it has decreased in high initial oil concentrations in the case of nanocomposites with 2 and 4 wt % of cloisite 20A. The nanocomposites with 3 wt % improve the sorption capacity by up to 16% and oil removal efficiency by up to 56% in water–oil system.[117] Magnetic polystyrene-palygorskite nanocomposites were prepared by a heterogeneous phase polymerization. The coating of polystyrene to the organo-$Fe_3O_4$-palygorskite makes it a water-repellent and oil-absorbing surface. Removal of oil from the surface of a nanocomposite was done by applying an external magnetic field. The oil removal capability test results indicated the removal of approximately 98% of oil.[118] The spongy at-tapulgite (ATP) was synthesized by themolysis of ATP/polymers nanocomposites are a good natural clay-based absorbent for oil. The treatment with polytetrafluoroethylene and polydime-thylsiloxane makes it superoleophilic to oil.[119]

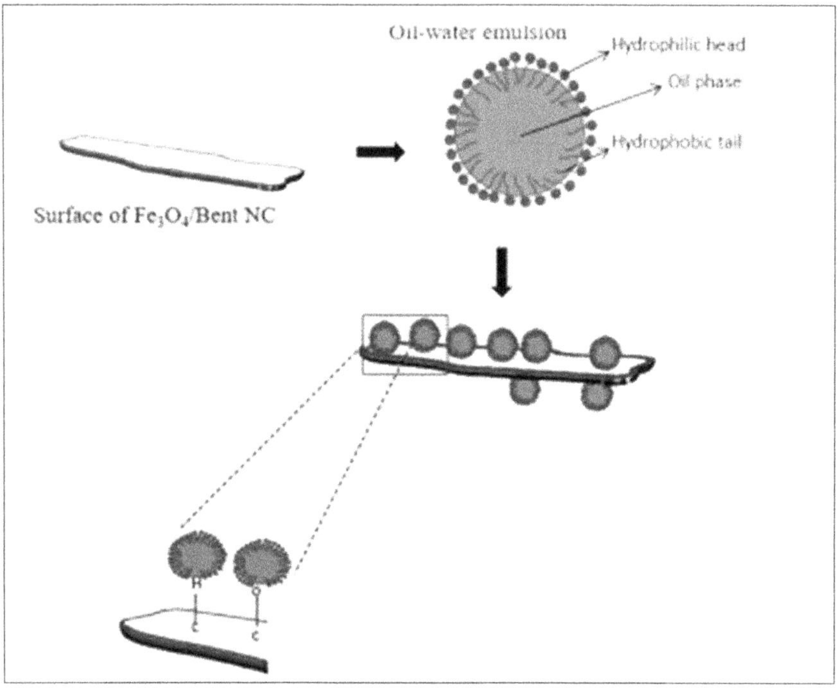


**FIGURE 4.10** The possible adsorption mechanism of diesel oil by $Fe_3O_4$/Bent NC.

*Source:* Reproduced with permission from Ref. [116]. Copyright © 2020 The Authors. Published by Elsevier Ltd. https://creativecommons.org/licenses/by/4.0/

#### *4.3.4.2 REMOVAL OF OIL BY METAL/METAL OXIDE-BASED POLYMER NANOCOMPOSITES*

Bolvardi et al. applied titanium dioxide ($TiO_2$) nanoparticles and polydimethylsiloxane (PDMS) on metal meshes with different pore sizes. Oil–water separation studies showed that the mesh with a smaller pore size exhibited a more efficient separation while a much shorter separation time and higher flux could be achieved by the mesh with a larger pore size.[120] $Fe_3O_4$ particles bound to the polyurethane sponge's skeleton through the self-polymerization of dopamine are suitable for the separation of oil.[121] Sharma et al. synthesized a magnetic nanosorbent polydimethylsiloxane @zinc oxide tetrapod@iron oxide nanorod nanohybrid (PZF) by hydrothermal process. The prepared nanohybrid exhibits exceptional adsorption ability and magnetic separation due to its large surface area and magnetic responsive properties, making it an alternative strategy for oily wastewater treatment.[122]

#### *4.3.4.3 REMOVAL OF OIL BY CARBON-BASED POLYMER NANOCOMPOSITES*

Carbon is an atom with a unique electronic structure that enables it to make covalent bonds with a wide range of other metals and nonmetals. Compared to other adsorbent materials, carbon at the nanoscale has many unique properties such as high mechanical properties, high electrical conductivity, high thermal stability, and so on. The excellent properties such as high aqueous solubility, large surface area, and enhanced thermal and mechanical stability of the polymer composites of carbon materials make it more attention in different applications, especially in wastewater treatment.

The SWCNTs were added to a copolymer membrane of poly(N- isopropyl acrylamide)-)-co-(acrylamide) (pNIPAm-co-AAm) for the separation of the water–oil mixture due to its oleophilic property.[123] This SWCNT-based membrane showed a 99% removal of oil. Its photothermal response characteristic was observed to boost the flow of water when illuminated with light.

The MWCNTs were incorporated into a block copolymer of polysulfone/polyether to form a CNTs-polymer membrane that separates oil from a water–oil mixture. The increase of MWCNTs up to 2.0 wt % has enhanced the oil rejection of the membrane from 91.4 to 99.79%. When the MWCNTs loading ratio is reached 0.5 wt %, the permeation flux has increased three times.[124] The polystyrene (PS)/MWCNT nanocomposite membranes containing various concentrations of CNT are fabricated by the electrospinning technique. It was irradiated with gamma radiation to induce enough crosslinks to enhance the polymer-filler interactions. The membrane shows superior hydrophobicity and selective oil absorption at 0.5 wt % CNT concentration.[125] An ultrathin and flexible film based on an SWCNT/$TiO_2$ nanocomposite was prepared by coating TiO2 via the sol–gel process onto an SWCNT ultrathin network film. The irradiation of UV light enhances the super hydrophilic and underwater superoleophobic properties. The high water flux value shows that it can effectively separate both surfactant-free and surfactant-stabilized oil-in-water emulsions with a wide range of droplet sizes.[126]

Alammar et al. developed nanocomposite membranes comprising polybenzimidazole (PBI), GO, and reduced GO (rGO) using phase inversion and common blade coating techniques. This membrane was dip-coated by polydopamine (PDA), which is known for its antifouling properties. The introduction of 0.5–1.5 wt % GO into the polymer matrix enhanced the performance and resulted in superior oil-removal efficiency of up to 99.9%. The mechanical stability of the membrane is also increased by GO.[127] The melamine sponge (MS) coated with rGO sheets exhibited a higher absorption capacity of 41 to 91 times for various oils due to its surface topography and high porosity.[128] The outstanding absorption capacity and excellent oil-water separation efficiency of carbon-based PNC, make it a promising candidate for oil–water separation.

### *4.3.5 DESALINATION*

Water is the basic requirement for creatures to survive. Out of 1.4 billion $km^3$ of total water supply on the planet, 97.5% of that is salt water, which cannot be used for human use. Therefore, one of the most significant commercial procedures to supply fresh water is the desalination of seawater. In the desalination process, saline water is split into two parts: fresh water, which has a low concentration of dissolved salts; and brine concentrate, which typically has a much higher concentration of dissolved salts than the original feed water.[129] There are two types of desalination techniques—thermal desalination and membrane desalination. The location of the desalination plant, the ease of obtaining the necessary energy, the cost of energy, etc., all play a role in determining the best desalination technique.[130] Due to improved fouling propensity, permeability, and salt rejection, polymeric-nanocomposite membranes (PNCMs) have captured the attention in performing desalination of saline and brackish water.[131] The various PNCs that can be applied in the desalination process involve clay-based PNCs, carbon-based PNCs, metal/metal oxide-based PNCs, etc. Polymers including polyvinylidene fluoride (PVDF), polyether sulfone (PES), polyacrylonitrile (PAN), PVA, polyvinyl chloride (PVC), polyethylene (PE), polypropylene (PP), polyamide, and chitosan are among the materials used in such membranes and composites.[132,133]

### *4.3.5.1 CLAY-BASED POLYMER NANOCOMPOSITES FOR DESALINATION*

Halloysites ($Al_2Si_2O_5(OH)_4{\cdot}nH_2O$) are natural clay minerals used for water separation due to their internal and exterior ionic structure, as well as the hollow tube form. In several studies, halloysite nanotubes (HNT) were used as filler in polymers to produce thin-film nanocomposite (TFN) membranes for seawater purification. For example, halloysite in various concentrations (from 0 to 5% by weight) was used as a filler in polylactic acid to form TFN membrane, and the results indicated that 5 wt % HNT-loaded membranes provided the best desalination performance with improved stability and hydrophilicity for the membrane. The greatest flow of 13.14 $kg/m^2.h$ was obtained, with a salt rejection rate of 99.95%.[134] In another example, TFN-forward osmosis (FO) membranes were prepared by using HNT as filler in polyamide. Because of its high-water permeability and low reverse solute flux, TFN membrane embedded with 0.05% HNTs was identified as the best-performing membrane.[135]

A novel polysulfone (poly-AF) incorporated with sepiolites nanorods-nanocomposite (poly-AF/Sep) was developed, which uses reverse osmosis for desalination. As a consequence, including optimized quantities of sepiolite (5 wt %) into the membrane matrix increased membrane performance, with a maximum water flux of 40.0 $Lm^{-2}h^{-1}$ and NaCl rejection of 99.1% at 30 bar pressure.[136] Using direct contact membrane distillation (DCMD) to desalinate seawater and produce potable water, highly hydrophobic poly (vinylidene fluoride) and clay nanocomposite were used to produce electrospun nanofiber membranes. The salt rejection was found to be higher than 99% for membranes with higher amounts of clay blended in and it prevents pore wetting in the DCMD process.[137]

### *4.3.5.2 METAL AND METAL OXIDE-BASED POLYMER NANOCOMPOSITES FOR DESALINATION*

Modifying the membrane surface using suitable metal and metal oxide nanoparticles has a significant effect on desalination performance as the permeability, selectivity, hydrophilicity, conductivity, thermal stability, mechanical strength, and antiviral and antibacterial properties of polymeric membranes are all affected by nanoparticles.[138,139]

Xiao et al. came up with a TFN membrane for the desalination process by blending porous Zr-based metal-organic frameworks (Zr-MOFs) into polyamide. It was found that the amount of porous material does not affect how much salt polyamide membranes reject by utilizing the pore adsorption reaction monomer principle.[140] In another research, by applying "defective ligand" strategy, a TFN membrane integrated with water-stable zirconium metal−organic cages was fabricated. This TFN membrane demonstrated 4.0 times more water flux than conventional polyamide thin-film composite (TFC) membrane despite sacrificing salt rejection.[141]

Porous Zn-based metal−organic frameworks (Zn-MOF)-incorporated TFN membrane were used to enhance desalination as well as chlorine resistance. The best-performing membrane (50 mg/L Zn-MOF nanofiller added) maintained benchmark selectivity to reject NaCl (>90%) and Na2SO4 (>95%), along with high water permeability of $2.46 \pm 0.12$ LMH/bar.[142]

Ultrafiltration-mixed matrix membranes (MMMs) formed by the incorporation of hydrophilic titania-based $TiO_2$-$ZrO_2$ (TiZr) and $TiO_2$-ZnO (TiZn) into a sulfonated poly(ether ketone)-polyethersulfone (SPEEK/PES) matrix demonstrated increased antifouling properties for seawater desalination. About 0.5 wt % TiZn-loaded SPEEK/PES MMMs with a higher water recovery percentage demonstrated a maximum flow of 88 $L/m^2$ h for seawater filtration. With an increase in TiZr and TiZn loading from 0.25 to 0.50 wt %, the hydrophilic behavior and water permeability of the resulting membranes improved.[143]

### *4.3.5.3 CARBON-BASED POLYMER NANOCOMPOSITES FOR DESALINATION*

MWCNTs are employed as efficient modifiers for boosting TFN desalination efficacy due to their high porosity (surface area) resulting from their nanoscale size.[144] A reverse osmosis membrane was developed using two types of raw and oxidized MWCNTs embedded in the polyamide layer. The water flux for the oxidized MWCNTs membranes was 28.9 $L/m^2$ h at 0.005 wt % and 25.9 $L/m^2$ h for raw MWCNTs at the same concentration.[145]

Polypyrrole (PPy)-coated MWCNTs were used to fabricate novel polyamide-nanocomposite reverse osmosis (RO) membranes, which showed an increase in the efficiency for desalination with increased water flux and antifouling properties. In comparison to MWCNTs membranes without polypyrrole structure, the newly modified MWCNTs-PPy membranes with hydrogen bonds provide hydrophilic surfaces, improving membrane performance.[146] Likewise, polyamide (PA) TFN incorporated with acid-treated MWCNT-titania nanotube hybrid (MWCNT-TNT) showed enhanced desalination without compromising the performance of salt rejection and water permeability greatly enhanced. The TFN membrane with 0.05 wt % acid-treated MWCNT-TNT had the maximum water permeability of 0.74 L/m2.h bar, roughly 57.45% higher than the plain PA membrane.[147] A TFN-nanofiltration (NF) membrane using amine-functionalized MWCNT ($NH_2$-MWCNT) as a hydrophilic modifier embedded in polyamide was developed to improve desalination performance. The separation performance of the NF membrane modified with 0.005 wt % $NH_2$-MWCNT was the best, with 36.71 and 95.72% rejection of $Na_2SO_4$ and NaCl salts, respectively.[148] It was reported that zwitterion functionalized carbon nanotube (CNT)/polyamide nanocomposite membranes showed an improved ion rejection ratio of 98.6%. The partial blocking of the tube ends by the functional groups and electrostatic repulsion between them and the ions, with steric effects dominating, are the two main explanations for the increase in ion rejection for the zwitterion functionalized CNTs.[149]

Hydrophobic membranes are promising candidates used in membrane distillation (MD) for efficient brine treatment. An example is polyvinylidene (PVDF)/graphene composite membrane, which showed the best pore size (0.56 m), porosity (69%), and hydrophobicity (139.5°) characteristics to be employed in water desalination in the MD process when 0.5 wt % graphene nanofiller was loaded. These results suggest that well-dispersed graphene nanoplatelets in a PVDF membrane matrix have significant applications in the desalination and microbial disinfection of actual seawater employing MD methods.[150]

## 4.4 CONCLUSION

Sustainable polymer nanocomposites offer a promising avenue for water purification due to their enhanced mechanical, thermal, and chemical properties. These materials combine the advantages of polymers with the unique characteristics of nanoscale fillers, resulting in improved performance in terms of water treatment. Various types of PNCs can be used for water treatments such as removal of HMIs, dye removal, oil pollutants removal and desalination. Sustainable PNCs hold great potential for addressing water purification challenges, and ongoing research in this field aims to develop effective and environmentally friendly solutions.

## KEYWORDS


- **polymer nanocomposites**
- **water purification**
- **clay**
- **metal and metal oxides**


## REFERENCES


1. Vaia, R. A.; Giannelis, E. P. Polymer Nanocomposites: Status and Opportunities. *MRS Bull.* **2001,** *26* (5), 394–401.
2. Kamal, A.; Ashmawy, M.; Algazzar, A. M.; Elsheikh, A. H. Fabrication Techniques of Polymeric Nanocomposites: A Comprehensive Review. *Proc. Inst. Mech. Eng. Part C: J. Mech. Eng. Sci.* **2022,** *236* (9), 4843–4861.
3. Pathak, M.; Slade, R.; Pichs-Madruga, R.; Ürge-Vorsatz, D.; Shukla, R.; Skea, J. Climate Change 2022 Mitigation of Climate Change: Technical Summary, 2022.
4. McDonald, R. I.; Green, P.; Balk, D.; Fekete, B. M.; Revenga, C.; Todd, M.; Montgomery, M. Urban Growth, Climate Change, and Freshwater Availability. *Proc. Natl. Acad. Sci.* **2011,** 108 (15), 6312–6317.
5. Gosling, S. N.; Arnell, N. W. A Global Assessment of the Impact of Climate Change on Water Scarcity. *Clim. Change* **2016,** *134*, 371–385.
6. Shannon, M. A.; Bohn, P. W.; Elimelech, M.; Georgiadis, J. G.; Mariñas, B. J.; Mayes, A. M. Science and Technology for Water Purification in the Coming Decades. *Nature* **2008,** *452* (7185), 301–310.
7. Subramani, A.; Jacangelo, J. G. Emerging Desalination Technologies for Water Treatment: A Critical Review. *Water Res.* **2015,** *75*, 164–187.
8. Faust, S. D.; Aly, O. M. *Chemistry of Water Treatment*; CRC Press, 2018.
9. Rashid, R.; Shafiq, I.; Akhter, P.; Iqbal, M. J.; Hussain, M. A State-of-the-Art Review on Wastewater Treatment Techniques: The Effectiveness of Adsorption Method. *Environ. Sci. Pollut. Res.* **2021,** *28*, 9050–9066.
10. Faust, S. D.; Aly, O. M. *Adsorption Processes for Water Treatment*; Elsevier, 2013.

11. Bolisetty, S.; Peydayesh, M.; Mezzenga, R. Sustainable Technologies for Water Purification from Heavy Metals: Review and Analysis. *Chem. Soc. Rev.* **2019,** *48* (2), 463–487.
12. Rekhate, C. V.; Srivastava, J. K. Recent Advances in Ozone-Based Advanced Oxidation Processes for Treatment of Wastewater—A Review. *Chem. Eng. J. Adv.* **2020,** *3*, 100031.
13. Nguyen, H. T.; Yoon, Y.; Ngo, H. H.; Jang, A. The Application of Microalgae in Removing Organic Micropollutants in Wastewater. *Crit. Rev. Environ. Sci. Technol.* **2021,** *51* (12), 1187–1220.
14. Pandey, N.; Shukla, S. K.; Singh, N. B. Water Purification by Polymer Nanocomposites: An Overview. *Nanocomposites* **2017,** *3* (2), 47–66.
15. Srivastava, N. K.; Majumder, C. B. Novel Biofiltration Methods for the Treatment of Heavy Metals from Industrial Wastewater. *J. Hazard. Mater.* **2008,** *151* (1), 1–8.
16. Bulut, Y. Removal of Heavy Metals from Aqueous Solution by Sawdust Adsorption. *J. Environ. Sci.* **2007,** *19* (2), 160–166.
17. Zare, E. N.; Motahari, A.; Sillanpää, M. Nanoadsorbents Based on Conducting Polymer Nanocomposites with Main Focus on Polyaniline and Its Derivatives for Removal of Heavy Metal Ions/Dyes: A Review. *Environ. Res.* **2018,** *162*, 173–195.
18. Lau, W. J.; Emadzadeh, D.; Shahrin, S.; Goh, P. S.; Ismail, A. F. Ultrafiltration Membranes Incorporated with Carbon-Based Nanomaterials for Antifouling Improvement and Heavy Metal Removal. In: *Carbon-Based Polymer Nanocomposites for Environmental and Energy Applications*; Elsevier, 2018; pp 217–232.
19. Hnamte, M.; Pulikkal, A. K. Clay-Polymer Nanocomposites for Water and Wastewater Treatment: A Comprehensive. *Chemosphere* **2022,** 135869.
20. Hızal, J.; Yılmazoğlu, M. Montmorillonite Clay Composite for Heavy Metal Removal from Water. In: *Green Adsorbents to Remove Metals, Dyes and Boron from Polluted Water*; 2021, pp 93–112.
21. Zeng, Z. L.; Yu, C.; Liao, R. P.; Cai, X.; Chen, Z. H.; Yu, Z.; Wu, Z. X. Preparation and Characterization of Sodium Polyacrylate Grafted Montmorillonite Nanocomposite for the Adsorption of Cadmium Ions form Aqueous Solution. *Colloids Surf. A: Physicochem. Eng. Aspects* **2023,** *656*, 130389.
22. Fayazi, M.; Ghanbarian, M. One-Pot Hydrothermal Synthesis of Polyethylenimine Functionalized Magnetic Clay for Efficient Removal of Noxious Cr (VI) from Aqueous Solutions. *Silicon* **2020,** *12* (1), 125–34.
23. Hussain, D.; Khan, S. A.; Alharthi, S. S.; Khan, T. A. Insight into the Performance of Novel Kaolinite-Cellulose/Cobalt Oxide Nanocomposite as Green Adsorbent for Liquid Phase Abatement of Heavy Metal Ions: Modelling and Mechanism. *Arab. J. Chem.* **2022,** *15* (7), 103925.
24. He, G.; Wang, C.; Cao, J.; Fan, L.; Zhao, S.; Chai, Y. Carboxymethyl Chitosan-Kaolinite Composite Hydrogel for Efficient Copper Ions Trapping. *J. Environ. Chem. Eng.* **2019,** *7* (2), 102953.
25. Chávez-Guajardo, A. E.; Medina-Llamas, J. C.; Maqueira, L.; Andrade, C. A.; Alves, K. G.; de Melo, C. P. Efficient Removal of Cr (VI) and Cu (II) Ions from Aqueous Media by Use of Polypyrrole/Maghemite and Polyaniline/Maghemite Magnetic Nanocomposites. *Chem. Eng. J.* **2015,** *281*, 826–836.

26. Agrawal, S.; Singh, N. B. Removal of Toxic Hexavalent Chromium from Aqueous Solution by Nickel Ferrite-Polyaniline Nanocomposite. *Desalin. Water Treatment* **2016,** *57* (38), 17757–17766.
27. Sahu, S.; Sahu, U. K.; Patel, R. K. Modified Thorium Oxide Polyaniline Core–Shell Nanocomposite and Its Application for the Efficient Removal of Cr (VI). *J. Chem. Eng. Data* **2019,** *64* (3), 1294–1304.
28. Bhaumik, M.; Noubactep, C.; Gupta, V. K.; McCrindle, R. I.; Maity, A. Polyaniline/Fe Composite Nanofibers: An Excellent Adsorbent for the Removal of Arsenic from Aqueous Solutions. *Chem. Eng. J.* **2015,** *271*, 135–146.
29. Barkade, S. S.; Nakate, U. T.; Bansod, P. G.; Deorukhkar, O. A.; Doss, V. R. Ultrasound-Assisted Synthesis of Polypyrrole/$TiO_2$ Nanocomposite for Cu (II) Ions Removal from Aqueous Solution. *Synth. Metals* **2023,** *293*, 117253.
30. Behbahani, M.; Babapour, M.; Amini, M. M.; Sadeghi, O.; Bagheri, A.; Salarian, M.; Rafiee, B. Separation/Enrichment of Copper and Silver Using Titanium Dioxide Nanoparticles Coated with Poly-Thiophene and Their Analysis by Flame Atomic Absorption Spectrophotometry. *Am. J. Analyt. Chem.* **2013,** *4*, 90–98.
31. Herab, A. A.; Salari, D.; Ostadrahimi, A.; Olad, A. Preparation of Magnetic Inulin Nanocomposite and Its Application in the Removal of Methylene Blue and Heavy Metals from Aqueous Solution. *Mater. Chem. Phys.* **2022,** *291*, 126580.
32. Sarojini, G.; Venkateshbabu, S.; Rajasimman, M. Facile Synthesis and Characterization of Polypyrrole-Iron Oxide–Seaweed (PPy-$Fe_3O_4$-SW) Nanocomposite and Its Exploration for Adsorptive Removal of Pb (II) from Heavy Metal Bearing Water. *Chemosphere* **2021,** *278*, 130400.
33. Pu, S.; Hou, Y.; Yan, C.; Ma, H.; Huang, H.; Shi, Q.; Mandal, S.; Diao, Z.; Chu, W. In Situ Coprecipitation Formed Highly Water-Dispersible Magnetic Chitosan Nanopowder for Removal of Heavy Metals and Its Adsorption Mechanism. *ACS Sustain. Chem. Eng.* **2018,** *6* (12), 16754–16765.
34. Liu, Z.; Gao, Z.; Xu, L.; Hu, F. Polypyrrole Modified Magnetic Reduced Graphene Oxide Composites: Synthesis, Characterization and Application for Selective Lead Adsorption. *RSC Adv.* **2020,** *10* (30), 17524–17533.
35. Sun, Y.; Li, D.; Yang, H.; Guo, X. Fabrication of $Fe_3O_4$@Polydopamine@Polyamidoamine Core–Shell Nanocomposites and Their Application for Cu (ii) Adsorption. *New J. Chem.* **2018,** *42* (14), 12212–12221.
36. Hong, J.; Xie, J.; Mirshahghassemi, S.; Lead, J. Metal (Cd, Cr, Ni, Pb) Removal from Environmentally Relevant Waters Using Polyvinylpyrrolidone-Coated Magnetite Nanoparticles. *RSC Adv.* **2020,** *10* (6), 3266–3276.
37. Saravaia, H.; Gupta, H.; Popat, P.; Sodha, P.; Kulshrestha, V. Single-Step Synthesis of Magnesium-Doped Lithium Manganese Oxide Nanosorbent and Their Polymer Composite Beads for Selective Heavy Metal Removal. *ACS Appl. Mater. Interfaces* **2018,** *10* (50), 44059–44070.
38. Zhu, Y.; Murali, S.; Cai, W.; Li, X.; Suk, J. W.; Potts, J. R.; Ruoff, R. S. Graphene and Graphene Oxide: Synthesis, Properties, and Applications. *Adv. Mater.* **2010,** *22* (35), 3906–3924.
39. Kaleekkal, N. J.; Thanigaivelan, A.; Rana, D.; Mohan, D. Studies on Carboxylated Graphene Oxide Incorporated Polyetherimide Mixed Matrix Ultrafiltration Membranes. *Mater. Chem. Phys.* **2017,** *186*, 146–158.

40. Arshad, F.; Selvaraj, M.; Zain, J.; Banat, F.; Haija, M. A. Polyethylenimine Modified Graphene Oxide Hydrogel Composite as an Efficient Adsorbent for Heavy Metal Ions. *Sep. Purif. Technol.* **2019,** *209*, 870–880.
41. Sambaza, S. S.; Masheane, M. L.; Malinga, S. P.; Nxumalo, E. N.; Mhlanga, S. D. Polyethyleneimine-Carbon Nanotube Polymeric Nanocomposite Adsorbents for the Removal of Cr from Water. *Phys. Chem. Earth, Parts A/B/C* **2017,** *100*, 236–246.
42. Chen, B.; Yue, W.; Zhao, H.; Long, F.; Cao, Y.; Pan, X. Simultaneous Capture of Methyl Orange and Chromium (VI) from Complex Wastewater Using Polyethylenimine Cation Decorated Magnetic Carbon Nanotubes as a Recyclable Adsorbent. *RSC Adv.* **2019,** *9* (9), 4722–4734.
43. Ahmad, K. Study of Different Polymer Nanocomposites and Their Pollutant Removal Efficiency. *Polymer* **2021,** *217*, 123453.
44. Maheshwari, K.; Agrawal, M.; Gupta, A. B. Dye Pollution in Water and Wastewater. *Novel Mater. Dye-Contain. Wastewater Treatment* **2021,** 1–25.
45. Amari, A.; Mohammed Alzahrani, F.; Mohammedsaleh Katubi, K.; Salem Alsaiari, N.; Tahoon, M. A.; Ben Rebah, F. Clay-Polymer Nanocomposites: Preparations and Utilization for Pollutants Removal. *Materials* **2021,** *14* (6), 1365.
46. Unuabonah, E. I.; Taubert, A. Clay–Polymer Nanocomposites (CPNs), Adsorbents of the Future for Water Treatment. *Appl. Clay Sci.* **2014,** *99*, 83–92.
47. Shabtai, I. A.; Lynch, L. M.; Mishael, Y. G. Designing Clay-Polymer Nanocomposite Sorbents for Water Treatment: A Review and Meta-Analysis of the Past Decade. *Water Res.* **2021,** *188*, 116571.
48. Mukhopadhyay, R.; Bhaduri, D.; Sarkar, B.; Rusmin, R.; Hou, D.; Khanam, R.; Sarkar, S.; Biswas, J. K.; Vithanage, M.; Bhatnagar, A.; Ok, Y. S. Clay–Polymer Nanocomposites: Progress and Challenges for Use in Sustainable Water Treatment. *J. Hazard. Mater.* **2020,** *383*, 121125.
49. Olad, A.; Azhar, F. F. Eco-Friendly Biopolymer/Clay/Conducting Polymer Nanocomposite: Characterization and Its Application in Reactive Dye Removal. *Fibers Polym.* **2014,** *15*, 1321–1329.
50. Deb, A.; Debnath, A.; Saha, B. Sono-Assisted Enhanced Adsorption of Eriochrome Black-T Dye onto a Novel Polymeric Nanocomposite: Kinetic, Isotherm, and Response Surface Methodology Optimization. *J. Dispersion Sci. Technol.* **2021,** *42* (11), 1579–1592.
51. Chen, J.; Feng, J.; Yan, W. Influence of Metal Oxides on the Adsorption Characteristics of PPy/Metal Oxides for Methylene Blue. *J. Colloid Interface Sci.* **2016,** *475*, 26–35.
52. Oyewo, O. A.; Elemike, E. E.; Onwudiwe, D. C.; Onyango, M. S. Metal Oxide-Cellulose Nanocomposites for the Removal of Toxic Metals and Dyes from Wastewater. *Int. J. Biolo. Macromol.* **2020,** *164*, 2477–2496.
53. Lefatshe, K.; Muiva, C. M.; Kebaabetswe, L. P. Extraction of Nanocellulose and In Situ Casting of ZnO/Cellulose Nanocomposite with Enhanced Photocatalytic and Antibacterial Activity. *Carbohydr. Polym.* **2017,** *164*, 301–308.
54. Ali Khan, M.; Govindasamy, R.; Ahmad, A.; Siddiqui, M. R.; Alshareef, S. A.; Hakami, A. A.; Rafatullah, M. Carbon Based Polymeric Nanocomposites for Dye Adsorption: Synthesis, Characterization, and Application. *Polymers* **2021,** *13* (3), 419.
55. Abdullah, N.; Tajuddin, M. H.; Yusof, N. Carbon-Based Polymer Nanocomposites for Dye and Pigment Removal. In: *Carbon-Based Polymer Nanocomposites for Environmental and Energy Applications*; Elsevier, 2018; pp 305–329.

56. Ghaemi, N.; Madaeni, S. S.; Daraei, P.; Rajabi, H.; Shojaeimehr, T.; Rahimpour, F. et al. PES Mixed Matrix Nanofiltration Membrane Embedded with Polymer Wrapped MWCNT: Fabrication and Performance Optimization in Dye Removal by RSM. *J. Hazard. Mater.* **2015,** *298*, 111–121.
57. Rouhi, M.; Lakouraj, M. M.; Tashakkorian, H.; Hasantabar, V. Novel Carbon Based Bioactive Nanocomposites of Aniline/Indole Copolymer for Removal of Cationic Dyes from Aqueous Solution: Kinetics and Isotherms. *New J. Chem.* **2019,** *43* (5), 2400–2410.
58. Gehlot, P.; Daga, K.; Mehta, R. Adsorption Study of Dye Water Using Poly Vinyl Alcohol Coated Carbon Black as an Effective and Low Cost Adsorbent. *Int. J. Chem.* **2011,** *3* (3), 56.
59. Maruthapandi, M.; Luong, J. H.; Gedanken, A. Kinetic, Isotherm and Mechanism Studies of Organic Dye Adsorption on Poly (4, 4′-oxybisbenzenamine) and Copolymer of Poly (4, 4′-Oxybisbenzenamine-pyrrole) Macro-Nanoparticles Synthesized by Multifunctional Carbon Dots. *New J. Chem.* **2019,** *43* (4), 1926–1935.
60. Hnamte, M.; Pulikkal, A. K. Clay-Polymer Nanocomposites for Water and Wastewater Treatment: A Comprehensive. *Chemosphere* **2022,** 135869.
61. Zhou, N. L.; Liu, Y.; Li, L.; Meng, N.; Huang, Y. X.; Zhang, J.; Wei, S. H.; Shen, J. A New Nanocomposite Biomedical Material of polymer/Clay–Cts–Ag Nanocomposites. *Curr. Appl. Phys.* **2007,** *7*, 58–62.
62. . Undabeytia, T.; Posada, R.; Nir, S.; Galindo, I.; Laiz, L.; Saiz-Jimenez, C.; Morillo, E. Removal of Waterborne Microorganisms by Fitration Using Clay–Polymer Complexes. *J. Hazard. Mater.* **2014,** *279*, 190–196.
63. Meng, N.; Zhou, N.-L.; Zhang, S.-Q.; Shen, J. Synthesis and Antimicrobial Activities of Polymer/Montmorillonite–Chlorhexidine Acetate Nanocomposite Films. *Appl. Clay Sci.* **2009,** *42*, 667–670.
64. Liu, E.; Sarkar, B.; Wang, L.; Naidu, R. Copper-Complexed Clay/Poly-Acrylic Acid Composites: Extremely Efficient Adsorbents of Ammonia Gas. *Appl. Clay Sci.* **2016,** *121*, 154–161.
65. Hernández-Hernández, K. A.; Illescas, J.; Díaz-Nava, M. D. C.; Martínez-Gallegos, S.; MuroUrista, C.; Ortega-Aguilar, R. E.; Rodríguez-Alba, E.; Rivera, E. Preparation of Nanocomposites for the Removal of Phenolic Compounds from Aqueous Solutions. *Appl. Clay Sci.* **2018,** *157*, 212–217.
66. Zhu, J.; Wang, T.; Zhu, R.; Ge, F.; Wei, J.; Yuan, P.; He, H. Novel Polymer/Surfactant Modified Montmorillonite Hybrids and the Implications for the Treatment of Hydrophobic Organic Compounds in Wastewaters. *Appl. Clay Sci.* **2011,** *51*, 317–322.
67. Unuabonah, E. I.; Olu-Owolabi, B. I.; Adebowale, K. O.; Yang, L. Z. Removal of Lead and Cadmium from Aqueous Solution by Polyvinyl Alcohol-Modified Kaolinite Clay: A Novel Nanoclay Adsorbent. *Adsorpt. Sci. Technol.* **2008,** *26* (6), 383–405.
68. Ganigar, R.; Rytwo, G.; Gonen, Y.; Radian, A.; Mishael, Y. G. Polymer–Clay Nanocomposites for the Removal of Trichlorophenol and Trinitrophenol from Water. *Appl. Clay Sci.* **2010,** *49*, 311–316.
69. Wang, X.; Du, Y.; Luo, J.; Yang, J.; Wang, W.; Kennedy, J. F. A Novel Biopolymer/Rectorite Nanocomposite with Antimicrobial Activity. *Carbohydr. Polym.* **2009,** *77*, 449–456.

70. Sölener, M.; Tunalib, S.; Özcanc, A. S.; Özcanc, A.; Gedikbey, T. Adsorption Characteristics of Lead (II) Ions onto the Clay/Poly (Methoxyethyl)Acrylamide (PMEA) Composite from Aqueous Solutions. *Desalination* **2008,** *223*, 308–322.
71. Zhu, J.; Wang, T.; Zhu, R.; Ge, F.; Wei, J.; Yuan, P.; He, H. Novel Polymer/Surfactant Modified Montmorillonite Hybrids and the Implications for the Treatment of Hydrophobic Organic Compounds in Wastewaters. *Appl. Clay Sci.* **2011,** *51*, 317–322.
72. Wang, T.; Zhu, R. L.; Ge, F.; Zhu, J. X.; He, H. P.; Chen, W. X. Sorption of Phenol and Nitrobenzene in Water by CTMAB/CPAM Organobentonites. *Huanjing Kexue/ Environ. Sci.* **2010,** *31*, pp. 385–389.
73. Anirudhan, T. S.; Suchithra, P. S. Heavy Metals Uptake from Aqueous Solutions and Industrial Wastewaters by Humic Acid-Immobilized Polymer/Bentonite Composite: Kinetics and Equilibrium Modeling. *Chem. Eng. J.* **2010,** *156*, 146–156.
74. Mousavi, M.; Aalemzadeh, I.; Vosoughi, M. Use of Modified Bentonite for Phenolic Adsorption in Treatment of Olive Oil Mill Wastewater. *Iran. J. Sci Technol. Trans. B: Eng.* **2006,** *30.*
75. Swallow, K. C.; Hume, D. N.; Morel, F. M. Sorption of Copper and Lead by Hydrous Ferric Oxide. *Environ. Sci. Technol.* **1980,** *14* (11), 1326–1331.
76. Kawashima, M.; Tainaka, Y.; Hori, T.; Koyama, M.; Takamatsu, T. Phosphate Adsorption onto Hydrous Manganese (IV) Oxide in the Presence of Divalent Cations. *Water Res.* **1986,** *20* (4), 471–475.
77. Jia, K.; Pan, B. C.; Zhang, Q. R.; Zhang, W. M.; Jiang, P. J.; Hong, C. H. et al. Adsorption of Pb2+, Zn2+, and Cd2+ from Waters by Amorphous Titanium Phosphate. *J. Colloid Interface Sci.* **2008,** *318*, 160–166.
78. Yang, W.; Yu, Z.; Pan, B.; Lv, L.; Zhang, W. Simultaneous Organic/Inorganic Removal from Water Using a New Nanocomposite Adsorbent: A Case Study of P-Nitrophenol and Phosphate. *Chem. Eng. J.* **2015,** *268*, 399–407.
79. Gunawan, P.; Guan, C.; Song, X.; Zhang, Q.; Leong, S. S.; Tang, C.; Chen, Y.; Chan-Park, M. B.; Chang, M. W.; Wang, K.; Xu, R. Hollow Fiber Membrane Decorated with Ag/MWNTs: Toward Effective Water Disinfection and Biofouling Control. *ACS Nano* **2011,** *5* (12), 10033–10040.
80. Jain, P.; Pradeep, T. Potential of Silver Nanoparticle-Coated Polyurethane Foam as an Antibacterial Water Filter. *Biotechnol. Bioeng.* **2005,** *90* (1), 59–63.
81. Salam, M. A.; Obaid, A. Y.; El-Shishtawy, R. M.; Mohamed, S. A. Synthesis of Nanocomposites of Polypyrrole/Carbon Nanotubes/Silver Nano Particles and Their Application in Water Disinfection. *RSC Adv.* **2017,** *7* (27), 16878–16884.
82. Jin, S.; Fallgren, P. H.; Morris, J. M.; Chen, Q. Removal of Bacteria and Viruses from Waters Using Layered Double Hydroxide Nanocomposites. *Sci. Technol. Adv. Mater.* **2007,** *8* (1–2), 67.
83. Moslehyani, A.; Ismail, A. F.; Othman, M. H.; Matsuura, T. Hydrocarbon Degradation and Separation of Bilge Water via a Novel $TiO_2$-HNTs/PVDF-Based Photocatalytic Membrane Reactor (PMR). *RSC Adv.* **2015,** *5* (19), 14147–14155.
84. Munnawar, I.; Iqbal, S. S.; Anwar, M. N.; Batool, M.; Tariq, S.; Faitma, N.; Khan, A. L.; Khan, A. U.; Nazar, U.; Jamil, T.; Ahmad, N. M. Synergistic Effect of Chitosan-Zinc Oxide Hybrid Nanoparticles on Antibiofouling and Water Disinfection of Mixed Matrix Polyethersulfone Nanocomposite Membranes. *Carbohydr. Polym.* **2017,** *175*, 661–670.

85. Dehaghi, S. M.; Rahmanifar, B.; Moradi, A. M.; Azar, P. A. Removal of Permethrin Pesticide from Water by Chitosan–Zinc Oxide Nanoparticles Composite as an Adsorbent. *J. Saudi Chem. Soc.* **2014,** *18* (4), 348–355.
86. Al-Naamani, L.; Dobretsov, S.; Dutta, J.; Burgess, J. G. Chitosan-Zinc Oxide Nanocomposite Coatings for the Prevention of Marine Biofouling. *Chemosphere* **2017,** *168*, 408–417.
87. Crane, R. A.; Dickinson, M.; Popescu, I. C.; Scott, T. B. Magnetite and Zero-Valent Iron Nanoparticles for the Remediation of Uranium Contaminated Environmental Water. *Water Res.* **2011,** *45* (9), 2931–2942.
88. Sun, Y.; Ding, C.; Cheng, W.; Wang, X. Simultaneous Adsorption and Reduction of U (VI) on Reduced Graphene Oxide-Supported Nanoscale Zerovalent Iron. *J. Hazard. Mater.* **2014,** *280*, 399–408.
89. Jang, J. H. Surface Chemistry of Hydrous Ferric Oxide and Hematite as Based on Their Reactions With Fe (II) and U (VI), 2004.
90. Pierce, M. L.; Moore, C. B. Adsorption of Arsenite and Arsenate on Amorphous Iron Hydroxide. *Water Res.* **1982,** *16* (7), 1247–1253.
91. Liu, L.; Zhang, Y.; Li, C.; Cao, J.; He, E.; Wu, X.; Wang, F.; Wang, L. Facile Preparation PCL/Modified Nano ZnO Organic-Inorganic Composite and Its Application in Antibacterial Materials. *J. Polym. Res.* **2020,** *27*, 1–1.
92. Fadakar Sarkandi, A.; Montazer, M.; Harifi, T.; Mahmoudi Rad, M. Innovative Preparation of Bacterial Cellulose/Silver Nanocomposite Hydrogels: In Situ Green Synthesis, Characterization, and Antibacterial Properties. *J. Appl. Polym. Sci.* **2021,** *138* (6), 49824.
93. McKay, G.; Porter, J. F.; Prasad, G. R. The Removal of Dye Colours from Aqueous Solutions by Adsorption on Low-Cost Materials. *Water Air Soil Pollut.* **1999,** *114*, 423–438.
94. Nasrollahzadeh, M.; Sajjadi, M.; Iravani, S.; Varma, R. S. Carbon-Based Sustainable Nanomaterials for Water Treatment: State-of-Art and Future Perspectives. *Chemosphere* **2021,** *263*, 128005.
95. Upadhyayula, V. K.; Deng, S.; Mitchell, M. C.; Smith, G. B. Application of Carbon Nanotube Technology for Removal of Contaminants in Drinking Water: A Review. *Sci. Total Environ.* **2009,** *408* (1), 1–3.
96. Lu, C.; Su, F. Adsorption of Natural Organic Matter by Carbon Nanotubes. *Sep. Purif. Technol.* **2007,** *58* (1), 113–121.
97. Gadhave, A.; Waghmare, J. Removal of Heavy Metal Ions from Wastewater by Carbon Nanotubes (CNTs). *Int. J. Chem. Sci. Appl.* **2014,** *5* (2), 56–67.
98. Abbas, M. F.; Abbas, A. H. Clozapine-Induced Myocarditis in Rats: Role of L-Carnitine in Protection. *Egypt. J. Forensic Sci. Appl. Toxicol.* **2016,** *16* (1), 141–157.
99. Shawky, H. A.; Chae, S. R.; Lin, S.; Wiesner, M. R. Synthesis and Characterization of a Carbon Nanotube/Polymer Nanocomposite Membrane for Water Treatment. *Desalination* **2011,** *272* (1–3), 46–50.
100. Asadzadeh-Khaneghah, S.; Habibi-Yangjeh, A. g-C3N4/Carbon Dot-Based Nanocomposites Serve as Efficacious Photocatalysts for Environmental Purification and Energy Generation: A Review. *J. Clean. Product.* **2020,** *276*, 124319.

101. Zhang, H.; Ming, H.; Lian, S.; Huang, H.; Li, H.; Zhang, L.; Liu, Y.; Kang, Z.; Lee, S.-T. $Fe_2O_3$/Carbon Quantum Dots Complex Photocatalysts and Their Enhanced Photocatalytic Activity Under Visible Light. *Dalton Trans.* **2011,** *40*, 10822–10825.
102. Ji, L.; Chen, W.; Duan, L.; Zhu, D. Mechanisms for Strong Adsorption of Tetracycline to Carbon Nanotubes: A Comparative Study Using Activated Carbon and Graphite as Adsorbents. *Environ. Sci. Technol.* **2009,** *43*, 2322–2327.
103. Cong, Q.; Yuan, X.; Qu, J. A Review on the Removal of Antibiotics by Carbon Nanotubes. *Water Sci. Technol.* **2013,** *68*, 1679–1687
104. Aslan, S.; Loebick, C. Z.; Kang, S.; Elimelech, M.; Pfefferle, L. D.; Van Tassel, P. R. Antimicrobial Biomaterials Based on Carbon Nanotubes Dispersed in Poly (Lactic-Co-Glycolic Acid). *Nanoscale* **2010,** *2* (9), 1789–1794.
105. Aslan, S.; Deneufchatel, M.; Hashmi, S.; Li, N.; Pfefferle, L. D.; Elimelech, M. et al. Carbon Nanotube-Based Antimicrobial Biomaterials Formed via Layer-by-Layer Assembly with Polypeptides. *J. Colloid Interface Sci.* **2012,** *388* (1), 268–273.
106. Ahmed, F.; Santos, C. M.; Vergara, R. A. M. V.; Tria, M. C. R.; Advincula, R.; Rodrigues, D. F. Antimicrobial Applications of electroactive PVK-SWNT Nanocomposites. *Environ. Sci. Technol.* **2012,** *46* (3), 1804–1810.
107. Mejias Carpio, I. E.; Santos, C. M.; Wei, X.; Rodrigues, D. F. Toxicity of a Polymer-Graphene Oxide Composite Against Bacterial Planktonic Cells, Biofilms, and Mammalian Cells. *Nanoscale* **2012,** *4* (15), 4746–4756.
108. Muthulingam, S.; Bae, K. B.; Khan, R.; Lee, I.-H.; Uthirakumar, P. Carbon Quantum Dots Decorated N-Doped ZnO: Synthesis and Enhanced Photocatalytic Activity on UV, Visible and Daylight Sources with Suppressed Photocorrosion. *J. Environ. Chem. Eng.* **2016,** *4*, 1148–1155.
109. Zhang, Y.; Wu, C.; Zhou, X.; Wu, X.; Yang, Y.; Wu, H.; Guo, S.; Zhang, J. Graphene Quantum Dots/Gold Electrode and Its Application in Living Cell $H_2O_2$ Detection. *Nanoscale* **2013,** *5*, 1816–1819.
110. Chen, Y.; Lu, Q.; Yan, X.; Mo, Q.; Chen, Y.; Liu, B.; Teng, L.; Xiao, W.; Ge, L.; Wang, Q. Enhanced Photocatalytic Activity of the Carbon Quantum Dot-Modified BiOI Microsphere. *Nanosc. Res. Lett.* **2016,** *11*, 60.
111. Elanchezhiyan, S. S.; Prabhu, S. M.; Kim, Y.; Park, C. M. Lanthanum-Substituted Bimetallic Magnetic Materials Assembled Carboxylate-Rich Graphene Oxide Nanohybrids as Highly Efficient Adsorbent for Perfluorooctanoic Acid Adsorption from Aqueous Solutions. *Appl. Surf. Sci.* **2002,** *509*, 144716.
112. Karlapudi, A. P.; Venkateswarulu, T. C.; Tammineedi, J.; Kanumuri, L.; Ravuru, B. K.; Ramu Dirisala, V.; Kodali, V. P. Role of Biosurfactants in Bioremediation of Oil Pollution–A Review. *Petroleum* **2018,** *4* (3), 241–249.
113. Chu, Z.; Feng, Y.; Seeger, S. Oil/Water Separation with Selective Superantiwetting/Superwetting Surface Materials. *Angew. Chem. Int. Ed.* **2015,** *54* (8), 2328–2338.
114. Saleh, T. A.; Parthasarathy, P.; Irfan, M. Advanced Functional Polymer Nanocomposites and Their Use in Water Ultra-Purification. *Trends Environ. Analyt. Chem.* **2019,** *24*, e00067.
115. Divya, S.; Oh, T. H. Polymer Nanocomposite Membrane for Wastewater Treatment: A Critical Review. *Polymers* **2022,** *14* (9), 1732.

116. Ewis, D.; Benamor, A.; Ba-Abbad, M. M.; Nasser, M.; El-Naas, M.; Qiblawey, H. Removal of Oil Content from Oil-Water Emulsions Using Iron Oxide/Bentonite Nano Adsorbents. *J. Water Process Eng.* **2020,** *38*, 101583.
117. Nikkhah, A. A.; Zilouei, H.; Asadinezhad, A.; Keshavarz, A. Removal of Oil from Water Using Polyurethane Foam Modified with Nanoclay. *Chem. Eng. J.* **2015,** *262*, 278–285.
118. Middea, A.; Spinelli, L.; Souza Junior, F. G.; Neumann, R.; Fernandes, T.; Faulstich, F. R.; Gomes, O. Magnetic Polystyrene–Palygorskite Nanocomposite Obtained by Heterogeneous Phase Polymerization to Apply in the Treatment of Oily Waters. *J. Appl. Polym. Sci.* **2018,** *135* (15), 46162.
119. Zhu, Z.; Dai, J.; Liu, Y.; Sun, H.; Liang, W.; Li, A. Hydrophobic Spongy Attapulgite for Absorption of Organics and Oils from Water. *Destech Trans. Mater. Sci. Eng.* **2017**.
120. Bolvardi, B.; Seyfi, J.; Hejazi, I.; Otadi, M.; Khonakdar, H. A.; Davachi, S. M. Towards an Efficient and Durable Superhydrophobic Mesh Coated by PDMS/$TiO_2$ Nanocomposites for Oil/Water Separation. *Appl. Surf. Sci.* **2019,** *492*:862–870.
121. Lu, J.; Liu, X.; Zhang, T. C.; He, H.; Yuan, S. Magnetic Superhydrophobic Polyurethane Sponge Modified with Bioinspired Stearic Acid@ $Fe_3O_4$@ PDA Nanocomposites for Oil/Water Separation. *Colloids Surf. A: Physicochem. Eng. Aspects* **2021,** *624*, 126794.
122. Sharma, M.; Joshi, M.; Nigam, S.; Avasthi, D. K.; Adelung, R.; Srivastava, S. K.; Mishra, Y. K. Efficient Oil Removal from Wastewater Based on Polymer Coated Superhydrophobic Tetrapodal Magnetic Nanocomposite Adsorbent. *Appl. Mater. Today* **2019,** *17*, 130–141.
123. Hu, L.; Gao, S.; Ding, X.; Wang, D.; Jiang, J.; Jin, J.; Jiang, L. Photothermal-Responsive Single-Walled Carbon Nanotube-Based Ultrathin Membranes for On/Off Switchable Separation of Oil-in-Water Nanoemulsions. *ACS Nano* **2015,** *9* (5), 4835–4842.
124. Berber, M. R. Current Advances of Polymer Composites for Water Treatment and Desalination. *J. Chem.* **2020,** *2020*, 1–9.
125. Parangusan, H.; Ponnamma, D.; Hassan, M. K.; Adham, S.; Al-Maadeed, M. A. Designing Carbon Nanotube-Based Oil Absorbing Membranes from Gamma Irradiated and Electrospun Polystyrene Nanocomposites. *Materials* **2019,** *12* (5), 709.
126. Gao, S. J.; Shi, Z.; Zhang, W. B.; Zhang, F.; Jin, J. Photoinduced Superwetting Single-Walled Carbon Nanotube/$TiO_2$ Ultrathin Network Films for Ultrafast Separation of Oil-in-Water Emulsions. *ACS Nano* **2014,** *8* (6), 6344–6352.
127. Alammar, A.; Park, S. H.; Williams, C. J.; Derby, B.; Szekely, G. Oil-in-Water Separation with Graphene-Based Nanocomposite Membranes for Produced Water Treatment. *J. Membr. Sci.* **2020,** *603*, 118007.
128. Sun, S.; Tang, S.; Chang, X.; Wang, N.; Wang, D.; Liu, T.; Lei, Y.; Zhu, Y. A Bifunctional Melamine Sponge Decorated with Silver-Reduced Graphene Oxide Nanocomposite for Oil-Water Separation and Antibacterial Applications. *Appl. Surf. Sci.* **2019,** *473*, 1049–1061.
129. Ibrahim, A. G.; Rashad, A. M.; Dincer, I. Exergoeconomic Analysis for Cost Optimization of a Solar Distillation System. *Solar Energy* **2017,** *151*, 22–32.
130. Harandi, H. B.; Rahnama, M.; Javaran, E. J.; Asadi, A. Performance Optimization of a Multi Stage Flash Desalination Unit with Thermal Vapor Compression Using Genetic Algorithm. *Appl. Therm. Eng.* **2017,** *123*, 1106–1119.

131. Shakeri, A.; Salehi, H.; Rastgar, M. Antifouling Electrically Conductive Membrane for Forward Osmosis Prepared by Polyaniline/Graphene Nanocomposite. *J. Water Process Eng.* **2019,** *32*, 100932.
132. Drioli, E.; Giorno, L.; Fontananova, E., Eds. *Comprehensive Membrane Science and Engineering*; Elsevier, 2017.
133. Bassyouni, M.; Abdel-Aziz, M. H.; Zoromba, M. S.; Abdel-Hamid, S. M.; Drioli, E. A Review of Polymeric Nanocomposite Membranes for Water Purification. *J. Ind. Eng. Chem.* **2019,** *73*, 19–46.
134. Nigiz, F. U.; Karakoca, B. Halloysite Nanotube Doped Poly Lactic Acid Membrane Preparation and Seawater Desalination. *Appl. Clay Sci.* **2023,** *231*, 106721.
135. Ghanbari, M.; Emadzadeh, D.; Lau, W. J.; Lai, S. O.; Matsuura, T.; Ismail, A. F. Synthesis and Characterization of Novel Thin Film Nanocomposite (TFN) Membranes Embedded with Halloysite Nanotubes (HNTs) for Water Desalination. *Desalination* **2015,** *358*, 33–41.
136. Khodami, S.; Mehdipour-Ataei, S.; Babanzadeh, S. Preparation, Characterization, and Performance Evaluation of Sepiolite-Based Nanocomposite Membrane for Desalination. *J. Ind. Eng. Chem.* **2020,** *82*, 164–172.
137. Prince, J. A.; Singh, G.; Rana, D.; Matsuura, T.; Anbharasi, V.; Shanmugasundaram, T. S. Preparation and Characterization of Highly Hydrophobic Poly (Vinylidene Fluoride)–Clay Nanocomposite Nanofiber Membranes (PVDF–Clay NNMs) for Desalination Using Direct Contact Membrane Distillation. *J. Membr. Sci.* **2012,** *397*, 80–86.
138. Yu, S.; Zuo, X.; Bao, R.; Xu, X.; Wang, J.; Xu, J. Effect of $SiO_2$ Nanoparticle Addition on the Characteristics of a New Organic–Inorganic Hybrid Membrane. *Polymer* **2009,** *50* (2), 553–559.
139. Ng, L. Y.; Mohammad, A. W.; Leo, C. P.; Hilal, N. Polymeric Membranes Incorporated with Metal/Metal Oxide Nanoparticles: A Comprehensive Review. *Desalination* **2013,** *308*, 15–33.
140. Xiao, F.; Hu, X.; Chen, Y.; Zhang, Y. Porous Zr-Based Metal-Organic Frameworks (Zr-MOFs)-Incorporated Thin-Film Nanocomposite Membrane Toward Enhanced Desalination Performance. *ACS Appl. Mater. Interfaces* **2019,** *11* (50), 47390–47403.
141. Liu, G.; Zhang, X.; Di Yuan, Y.; Yuan, H.; Li, N.; Ying, Y.; Peh, S. B.; Wang, Y.; Cheng, Y.; Cai, Y.; Gu, Z. Thin-Film Nanocomposite Membranes Containing Water-Stable Zirconium Metal–Organic Cages for Desalination. *ACS Mater. Lett.* **2021,** *3* (3), 268–274.
142. Shukla, A. K.; Alam, J.; Alhoshan, M. S.; Ali, F. A.; Mishra, U.; Hamid, A. A. Thin-Film Nanocomposite Membrane Incorporated with Porous Zn-Based Metal–Organic Frameworks: Toward Enhancement of Desalination Performance and Chlorine Resistance. *ACS Appl. Mater. Interfaces* **2021,** *13* (24), 28818–28831.
143. Lukka Thuyavan, Y.; Arthanareeswaran, G.; Ismail, A. F.; Goh, P. S.; Shankar, M. V.; Ng, B. C.; Sathish Kumar, R.; Venkatesh, K. Binary Metal Oxides Incorporated Polyethersulfone Ultrafiltration Mixed Matrix Membranes for the Pretreatment of Seawater Desalination. *J. Appl. Polym. Sci.* **2021,** *138* (8), 49883.
144. Abidin, M. N.; Goh, P. S.; Ismail, A. F.; Othman, M. H.; Hasbullah, H.; Said, N.; Kadir, S. H.; Kamal, F.; Abdullah, M. S.; Ng, B. C. Antifouling Polyethersulfone

Hemodialysis Membranes Incorporated with Poly (Citric Acid) Polymerized Multi-Walled Carbon Nanotubes. *Mater. Sci. Eng.: C* **2016,** *68*, 540–550.
145. Farahbakhsh, J.; Delnavaz, M.; Vatanpour, V. Investigation of Raw and oxidized Multiwalled Carbon Nanotubes in Fabrication of Reverse Osmosis Polyamide Membranes for Improvement in Desalination and Antifouling Properties. *Desalination* **2017,** *410*, 1–9.
146. Farahbakhsh, J.; Delnavaz, M.; Vatanpour, V. Simulation and Characterization of Novel Reverse Osmosis Membrane Prepared by Blending Polypyrrole Coated Multiwalled Carbon Nanotubes for Brackish Water Desalination and Antifouling Properties Using Artificial Neural Networks. *J. Membr. Sci.* **2019,** *581*, 123–138.
147. Azelee, I. W.; Goh, P. S.; Lau, W. J.; Ismail, A. F.; Rezaei-DashtArzhandi, M.; Wong, K. C.; Subramaniam, M. N. Enhanced Desalination of Polyamide Thin Film Nanocomposite Incorporated with Acid Treated Multiwalled Carbon Nanotube-Titania Nanotube Hybrid. *Desalination* **2017,** *409*, 163–170.
148. Zarrabi, H.; Yekavalangi, M. E.; Vatanpour, V.; Shockravi, A.; Safarpour, M. Improvement in Desalination Performance of Thin Film Nanocomposite Nanofiltration Membrane Using Amine-Functionalized Multiwalled Carbon Nanotube. *Desalination* **2016,** *394*, 83–90.
149. Chan, W. F.; Chen, H. Y.; Surapathi, A.; Taylor, M. G.; Shao, X.; Marand, E.; Johnson, J. K. Zwitterion Functionalized Carbon Nanotube/Polyamide Nanocomposite Membranes for Water Desalination. *ACS Nano* **2013,** *7* (6), 5308–5319.
150. Gontarek-Castro, E.; Rybarczyk, M. K.; Castro-Muñoz, R.; Morales-Jiménez, M.; Barragán-Huerta, B.; Lieder, M. Characterization of PVDF/Graphene Nanocomposite Membranes for Water Desalination with Enhanced Antifungal Activity. *Water* **2021,** *13* (9), 1279

# CHAPTER 5

# Sustainable Biopolymer-Based Materials for Healthcare Applications

AISWARYA SATHIAN[1,4], SNEHA JAMES[2], PARVATHY M. V.[2], SREELAKSHMI K. S.[2], NAMITHA VIJAY[5], SHASHANKA K. PRASAD[5], SABU THOMAS[2,4], and JOSHY K. S.[2]

[1]*College of Science and Engineering, James Cook University, Townsville, QLD, Australia*

[2]*School of Nanoscience and Nanotechnology, Mahatma Gandhi University, Kottayam, Kerala, India*

[4]*International and Inter University Centre for Nanoscience and Nanotechnology, Mahatma Gandhi University, Kottayam, Kerala, India*

[5]*Department of Biotechnology and Bioinformatics, JSSAHER, Mysuru, Karnataka, India*

## ABSTRACT

Regenerative medicine stands out as a propitious discipline within healthcare, presenting the prospect of repairing and surrogating damaged or ailing tissues and organs. A pivotal strategy in regenerative medicine involves employing scaffolds to maneuver cell growth, foster tissue regeneration, and contribute to the reconstruction of organs. Among the diverse arrays of scaffolds, those derived from natural biopolymers have garnered

Polymers and Functional Materials for a Cleaner Environment and Human Health: Sustainable Technologies and New Research. Ruby Varghese, Ann Rose Abraham, Shrikaant Kulkarni, & A. K. Haghi (Eds.)

considerable attention due to their compatibility with living organisms, capacity for gradual breakdown, and ability to replicate the extracellular matrix (ECM) found in native tissues. This chapter offers a comprehensive overview of how natural biopolymer-based scaffolds are applied in clinical settings within regenerative medicine, emphasizing their potential to transform the treatment landscape for various medical conditions.

## 5.1 INTRODUCTION

Natural biopolymer-based scaffolds are known for their excellent biocompatibility and nontoxicity, which are critical properties for their successful use in biomedical applications. Here are some key aspects related to these properties: Any substance, whether derived from natural or synthetic sources, alone or in combination, can be applied in entirety or as a component in any tissue, organ, or bodily function to either sustain or improve an individual's quality of life. This applied substitute can be categorized as a biomaterial.[1] When crafting scaffolds for soft and hard tissues, biocompatibility stands out as a crucial factor. The effectiveness of a biomaterial is determined by its accomplishment of the intended medical purpose without causing any detrimental local or systemic effects on the recipient of the therapy. Natural biopolymer-based scaffolds are compatible with biological systems and can interact with cells and tissues without causing adverse reactions. They create a conducive milieu for cellular adhesion, proliferation, and differentiation, thereby fostering the process of tissue regeneration.[2,3] The various applications of biopolymers in a biomedical field have been shown in Figure 5.1.

Natural biopolymers that come from renewable resources, including plants or animals, are frequently nontoxic. They guarantee the safety of the scaffold within the body since they have no negative or toxic effects on cells or tissues. These structures consist of extended chains, characterized by covalently linked repetitions of amino acids, nucleotides, or monosaccharides. Frequently, these polymers incorporate biofunctional compounds to ensure bioactivity, biomimetic characteristics, and natural reconfiguration. Natural biopolymers have a limited immunogenic potential, especially those originating from human or animal origins. This indicates that when implanted in the body, they are less likely to elicit an immunological response or result in rejection. When compared to synthetic materials, natural biopolymer-based scaffolds have less of a potential to cause inflammation.

The ability to reduce the immune response and speed up the healing process makes this trait useful for tissue engineering applications.[2]

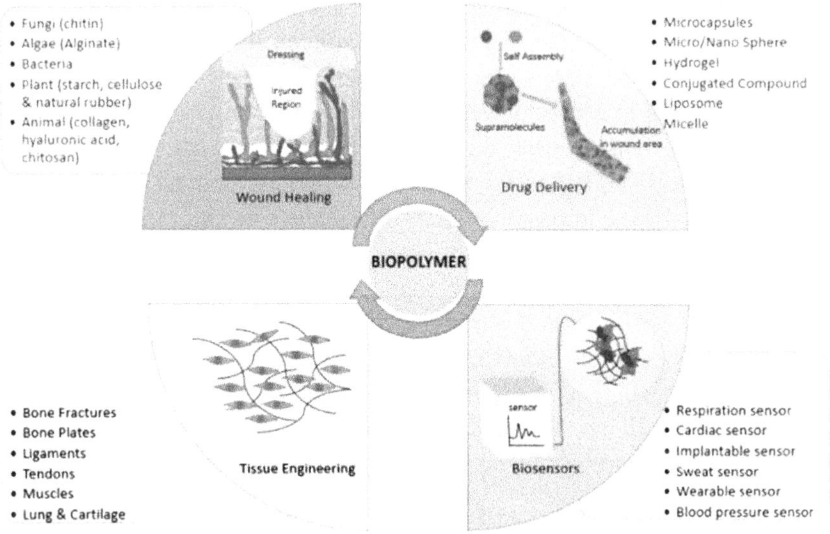


**FIGURE 5.1** Schematic representation of biopolymer in various biomedical fields.

*Source*: Reprinted with permission from Ref. [39]. Copyright © 2021, The Author(s), under exclusive licence to Springer Science Business Media, LLC, part of Springer Nature.

## 5.2 PROPERTIES OF NATURAL BIOPOLYMER-BASED SCAFFOLDS

The body's own enzymes often metabolize and break down natural biopolymers. This metabolic breakdown enhances the integration and remodeling of the regenerated tissue and decreases the build-up of scaffold residues. A lot of natural biopolymers are capable of being enzymatically broken down over time. This characteristic makes it possible for the scaffold to gradually deteriorate and change as a new tissue develops, negating the need for later surgical removal. Scaffolds can be taxonomically delineated into two categories predicated upon their degradation kinetics: permanent/nondegradable and biodegradable. They may degrade via mechanisms involving physical, chemical, or biological processes, as well as biological processes that use biological agents as intermediaries, such as tissue remodeling enzymes. Biodegradable polymers can be divided into two primary types based on how they were created: stepwise polycondensation or ring-opening polymerization. The first group of agro-polymers includes proteins and polysaccharides. The second group consists of biodegradable polyesters, such as aromatic and aliphatic polyesters. Biodegradable polymers that can replace nonbiodegradable polymers find applications across diverse domains, encompassing various applications, such as biomedical, textile, and packaging ones. They have a wide range of features.[2,3] Natural biopolymers derived from validated and well-characterized sources have a higher likelihood of receiving regulatory approval for biomedical applications. Their biocompatibility and nontoxicity profiles contribute to their acceptance as safe and effective materials in various medical contexts. The combination of biocompatibility and nontoxicity in natural biopolymer-based scaffolds ensures that they can interact harmoniously with the biological system, support tissue regeneration, and minimize potential risks associated with foreign-body reactions or toxicity.[4]

Natural biopolymer-based scaffolds offer a range of properties that contribute to their utility pertaining to the field of tissue engineering and regenerative medicine. In addition to biocompatibility and nontoxicity, here are additional key properties:

a) **Biodegradability and tunable degradation rates:** Natural biopolymers are often biodegradable, meaning they can be broken down by enzymatic or biological processes over time. The degradation rates of these scaffolds can be customized to align with the targeted timeline for tissue regeneration. This property enables the scaffold to gradually degrade as new tissue forms, providing support during the healing process.[2]
b) **Structural and mechanical properties:** Natural biopolymers possess inherent structural integrity and mechanical properties configured to perform specific functions. For instance, they can be engineered to have varying stiffness, elasticity, or strength to mimic the mechanical properties of the target tissue. This property ensures that the scaffold provides adequate support and facilitates proper cell function and tissue growth.
c) **Biomimetic Attributes Mimicking the Extracellular Matrix (ECM):** The extracellular matrix is a framework of proteins and polysaccharides that provides not only the physical framework for cells and tissues but also contains numerous instructive signals critical for tissue development, homeostasis, and repair. Natural biopolymer-based scaffolds can be designed to mimic the composition and structure of the ECM. These are emulated, establishing

a biomimetic microenvironment conducive to cellular activities. This promotes cell adhesion, migration, and differentiation, ultimately leading to the regeneration of functional tissue.[5,6]

d) **Porosity and interconnectivity:** Natural biopolymer-based scaffolds can be fabricated with a porous structure that allows for the diffusion of nutrients, oxygen, and waste products. The interconnected pore network facilitates cell infiltration, proliferation, and the formation of a new tissue. This property is crucial for maintaining cell viability and promoting tissue integration. The porosity of a scaffold depends on the number, size, shape, connection, and orientation of its pores. Porosity often encourages cell migration and proliferation, providing an environment for the delivery of nutrients to the structures beneath or close to the scaffolds. Porous scaffolds guarantee cellular proliferation, homogeneous cell distribution, and vascularization. Parameters such as average pore size, pore size distribution, pore length, pore interconnectivity, pore shape, pore throat size, and pore wall roughness are pivotal considerations in the construction process of a scaffold. As a temporary model for the growth and reorganization of the new tissue, it initiates the formation of a biocompatible, porous network, thereby inducing the surrounding tissue.[5,7,8]

e) **Bioactive properties:** Bioactivity refers to a material's ability to impact its biological surroundings. Some natural biopolymers possess bioactive properties that can further enhance tissue regeneration. For example, certain biopolymers can interact with growth factors, cytokines, or cell surface receptors, influencing cellular behavior and tissue development. Additionally, natural biopolymers may exhibit antimicrobial properties, helping to prevent infections during the healing process. Biomaterials may include biological indicators to strengthen adhesion, such as cell-adhesive ligands, or physical indications, such as topography, to affect the shape and orientation of cells.[2,9,10]

These properties collectively contribute to the effectiveness of natural biopolymer-based scaffolds in promoting tissue regeneration and functional tissue formation. By replicating key aspects of the native tissue microenvironment, these scaffolds provide a supportive framework for cell growth and differentiation, ultimately leading to successful tissue engineering.

## 5.3 CLINICAL APPLICATIONS OF NATURAL BIOPOLYMER-BASED SCAFFOLDS

Natural biopolymer-based scaffolds have a wide range of medical applications in tissue engineering and regenerative medicine. These scaffolds offer unique advantages, such as biocompatibility, biodegradability, and biomimetic properties, which make them suitable for various therapeutic interventions. Here are some clinical applications of natural biopolymer-based scaffolds: Collagen, silk, elastin, alginates, and fibers are examples of naturally occurring polymeric biomaterials that are used in tissue engineering because of their biocompatibility, bioactivity, and adjustable mechanical and degradation kinetics. These natural biopolymers are advantageous for use in biomedical applications because they do not produce cytotoxic degradation products, are frequently processed using environmentally friendly aqueous-based techniques, and can be tailored to control the rate of degradation in biological systems.[11,12]

### *5.3.1 BONE REGENERATION*

Bone regeneration is a promising field within tissue engineering that aims to restore damaged or lost bone tissue. Natural biopolymer-based scaffolds have gained significant attention in bone tissue engineering due to their biocompatibility, bioactivity, and ability to provide structural support and guide cellular behaviour.[12] Here are some key aspects related to the use of natural biopolymer scaffolds in bone regeneration, including case studies, clinical outcomes, challenges, and future prospects:

#### *5.3.1.1 USE OF NATURAL BIOPOLYMER SCAFFOLDS IN BONE TISSUE ENGINEERING*

Natural biopolymers, such as collagen, chitosan, gelatin, alginate, and hyaluronic acid, have been extensively investigated as scaffold materials for bone tissue engineering. These scaffolds can be fabricated into various forms, including hydrogels, porous scaffolds, and nanofibrous structures, to mimic the natural extracellular matrix (ECM) and provide an optimistic microenvironment for cell attachment, proliferation, and differentiation. These scaffolds can be combined with other components like bioactive ceramics (e.g., hydroxyapatite) or growth factors (e.g., bone) to enhance their osteo conductivity and osteo inductivity, respectively. Moreover, they can be functionalized with peptides or small molecules to promote specific cellular responses and improve bone formation.[11–13] Tissue engineering requires scaffolds with regulated 3D structures and gradients of physicochemical characteristics. Biopolymer aerogel scaffolds can be used to create these specialized structures. The creation of the required biocompatible aerogel scaffold structures using fast prototyping and 3D printing methods is possible. Remarkable advancements have been achieved in the regeneration of various human tissues, attributable to the distinctive capabilities inherent in biopolymers and the crystalline structure of aerogels.[14,15]

### *5.3.2 USE OF NATURAL BIOPOLYMER SCAFFOLDS IN CARTILAGE REPAIR*

A nonvascular flexible connective tissue called cartilage can be found in the space between two bones. It is a hard tissue with a matrix made of a gel-like material and 80% water. Poor joint alignment, excessive strain, being overweight, injuries, and other factors can all contribute to cartilage deterioration. The symptoms of cartilage damage are only momentarily relieved by current treatments, which do little to prevent cartilage degradation. Even while there are traditional surgical techniques including microfracture, autograft, and allograft mosaicplasty, they have drawbacks. Because of a lack of donor tissues, mosaicplasty frequently fails to produce articular hyaline cartilage. Additionally, many surgical procedures may necessitate additional surgeries and are linked to significant problems. A potential strategy for creating minimally invasive therapies for cartilage lesion repair is provided by biomedical engineering. The creation of one-step in situ cartilage repair therapies using biomaterials and endogenous cells is an emerging concept in biomaterial engineering. Among other endogenous cells, mesenchymal stem cells (BMSCs) from bone marrow are essential for effective cartilage repair. However, endogenous MSCs' restricted migratory and brief local retention currently limit their efficacy in cartilage repair.

### 5.3.2.1 NATURAL BIOPOLYMERS

Natural polymers demonstrate evident biological applicability in tissue regeneration owing to their inherent biocompatibility, biodegradability, and macromolecular similarity to the native extracellular matrices (ECMs), which can provide a fantastic bioactivity and natural sticky surface for cells essential for bioactivity. Materials based on proteins, such as gelatin, collagen, fibrin, and silk fibroin, as well as those derived from polysaccharides, such as hyaluronic acid, chondroitin sulfate, alginate, and others, constitute natural polymers employed in the fabrication of hydrogels. The secure elimination of ultimate metabolites from the organism is made possible by the fact that natural hydrogels cannot trigger immune or toxic reactions and that their degradation products are not toxic or immunogenic. However, their poor stability, fast disintegration, and relatively low mechanical strength severely restrict their applications. While amalgamating synthetic and natural materials is an effective and simple technique for integrating them, the advantages of creating hydrogels for the regeneration of cartilage tissue, the impermeability of the components, and the unpredictability of the metabolites have led to significant limitations in actual biomedicine. Reconstructions of cell morphology, growth, proliferation, differentiation, and new tissue formation are therefore hybridization of other natural polymers or advanced biomodification of natural hydrogels to acquire better mechanics. The types of natural biopolymers are shown in Figure 5.2.

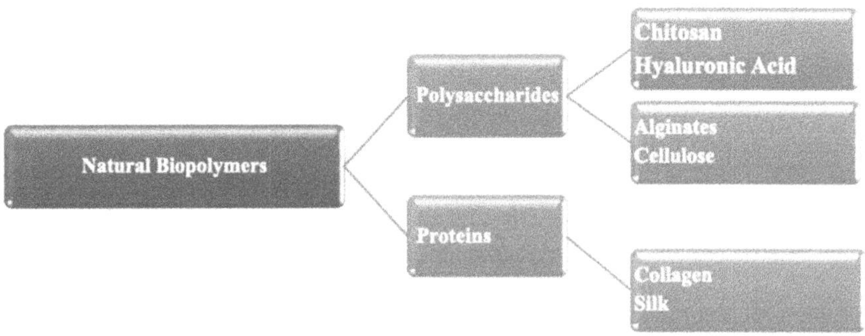


**FIGURE 5.2** Shows the types of natural biopolymers.

#### *5.3.2.1.1 Polysaccharide Based*

a) **Polysaccharides hyaluronic acid (HA)** assumes crucial functions in various physiological activities. Given its substantial presence within the cartilage extracellular matrix (ECM) and favorable biocompatibility, particularly with elevated molecular weight, HA is a frequent choice in the field of cartilage tissue engineering. Since HA, in particular, has biological properties, it is frequently utilized to create hydrogel scaffolds that interact with cells. Aggrecan and Sox9 expression in chondrocytes embedded in hydrogels containing 5% HA increased more than five times above that in hydrogels containing 1.5% HA.[18]

b) **Chitosan** is a cationic molecule because of the amino groups in it. It has strong bio-adhesive capabilities and antibacterial characteristics. Ampholytic chitosan/carrageenan hybrid hydrogels were created by Liang et al. and demonstrated high promise for medicinal applications for cartilage repair by inducing chondrogenic differentiation of ATDC5 cells in vitro. Chitosan was combined with demineralized bone matrix particles, modified with a peptide derived from phage display, to formulate a pliable hydrogel scaffold capable of addressing diverse types of cartilage defects.[19]

c) **Alginate** is commonly utilized as an injectable hydrogel through minimally invasive techniques for cartilage regeneration, alginate readily undergoes gelation with divalent cations, such as calcium ions (Ca2+). Given its swift cross-linking capability, alginate has also found applications in 3D bioprinting processes.

d) However, alginate is not bio-functional enough to interact with proteins and cells on its own. When HA and alginate were combined, Park et al. demonstrated that hybrid scaffolds could still form gels in the presence of Ca2+ and that ATDC5's chondrogenic differentiation was improved. Alginate hydrogels are interestingly capable of serving as gene carriers for tissue engineering. For example, combined MSCs with nHA (nanohydroxyapatite) and plasmid DNA (pDNA) encoding for pTGF-3, pBMP2, or both into alginate hydrogels. In contrast to the control without alginate, it was discovered that the pDNA was delivered to MSCs. The transfection of encapsulated MSCs was successfully supported by this unique gene-activated alginate hydrogel, which also successfully controlled the direction of the MSCs' orientation through several genes.

e) **Cellulose** is a linear chain polysaccharide made up of d-glucose units. Oftentimes, cellulose is altered to improve its characteristics before being used to create hydrogels. Gelatin and sodium cellulose sulfate (NaCS) can be used to create a fibrous scaffold in which hMSC chondrogenesis has been demonstrated to be improved by increased type II collagen gene expression. Using laponites and a novel isolated hydroxypropyl methylcellulose (Si-HPMC) hydrogel, a hydrogel scaffold with better mechanical strength, cytocompatibility, and oxygen diffusion was created. This Si-HPMC hydrogel created an ECM that was much more similar to cartilage when it was supplemented with a physiologically active marine exopolysaccharide. An optimistic environment for chondrocytes, facilitating cartilage formation both in vitro and in vivo, can be provided by a bilayer bacterial nanocellulose (BNC).[20]

Polysaccharides, being macromolecules with a highly biocompatible nature, bear a resemblance to the glycan component of the extracellular matrix (ECM). They have undergone extensive exploration within the realm of cartilage tissue engineering. Nevertheless, the identification of a singular type of polysaccharide remains elusive.

#### *5.3.2.1.2 Protein Based*

a) **Silk** is derived from a variety of Lepidoptera larvae, including silkworms and spiders. Silk naturally occurs in a variety of configurations, structures, and roles. Silk's source and environment cause this complicated behavior. The strongest natural fiber, silk fiber, is the ideal combination of high strength, low weight, outstanding durability, and suppleness. Silk is composed of two primary proteins: sericin, a glue-like protein with 18 different types of amino acids, and silk fibroin (SF), which is derived from the fibrous component of the filament. Studies have shown that the creation of silk fibroin-based scaffolds in a variety of topologies results in attractive biocompatibility in addition to a tenable breakdown rate and mechanical qualities. In contrast to alternative biological materials, fibroin has been evidenced to manifest predictable proteolytic degradation by modulating parameters such as its diameter, diverse failure intervals, failure strength, and mass degradation. In addition to its exceptional mechanical qualities, silk has biocompatibility, thermostability (up to 250°C), and a wide range of processing temperatures. The composition of the blended materials, the origin of silk, and the concentration levels of the materials in composites govern the silk quality.[21]

b) **Collagen** is an ECM protein that exists naturally in cartilage and aids in the formation of chondrocytes. As a result, it is frequently utilized in the engineering of cartilage tissue. By encouraging the development of the extracellular matrix (ECM), collagen hydrogels lower the immunogenicity of foreign cells seeded in designed hydrogel structures and encourage cartilage regeneration. Additionally, Wong et al. discovered that type II collagen may dedifferentiate auricular chondrocytes by monolayer passaging and then transform them into articular cartilage. Zhang et al. effectively stimulated the migration and adhesion of C-MSCs and SM-MSCs, facilitating the self-repair of partial-thickness defects. This was achieved by employing type I collagen as a scaffold in conjunction with stromal cell-derived factor-1 (SDF-1) as a growth factor.[22]

### *5.3.3 SKIN TISSUE ENGINEERING*

Skin is the largest organ in the human body and accounts for around one-tenth of total body mass. In general, it protects against osmotic, chemical, mechanical, and microbial invasion, and photodamage. Burns, soft tissue trauma, UV radiation, and chemical interaction can cause skin defects and necrosis. In tissue regeneration research, skin regeneration is of utmost importance. Skin defects can be repaired using traditional therapies such as autograft, allograft, and xenograft. Although autografts are frequently utilized to correct skin abnormalities, donor sites have drawbacks. As an alternative, allografts can increase income, but they frequently come with a substantial risk of immunological rejection and disease transmission. Tissue-engineered skin substitutes are employed to alleviate these issues with predictable treatments. These tissue-engineered skin substitutes' high prices and weak mechanical strength are two of their main drawbacks.

The body's outermost layer of defense against external infections and diseases is the skin. It contains two main layers: the outer layer, known as the epidermis, which serves as a barrier to the outside environment, and the inner layer, also known as the dermis, which is made up of connective tissue. Different types of skin injuries fall into the following categories: superficial (only the epidermis is harmed), partial thickness (most of the dermis is healthy), full thickness (the entire dermis, including the sweat glands and hair follicles, is damaged), and complex (all layers are injured). Acute wounds like burns, chemical burns, and surgical wounds heal in two to three months. Chronic wounds, such as those caused by diabetes, infections, and bedsores, however, do not heal normally. Loss of tissue results from the recurrence and delayed healing processes, which have an impact on the sublayers like muscles and bones. Chronic wounds can be brought on by any factor, such as systemic inflammatory disease, diabetes, and infections, that retards the transition from inflammation to proliferation. Loss of tissue results from the recurrence and delayed healing processes, which have an impact on the sublayers like muscles and bones. Chronic wounds can be brought on by any factor, such as systemic inflammatory disease, diabetes, and infections, that slows down the transition from inflammation to proliferation.

Diverse cellular activities, encompassing proliferation, migration, differentiation, and apoptosis, play pivotal roles in the process of skin wound healing. The conventional wound-healing process involves four interconnected stages: hemostasis, inflammation, proliferation, and remodeling.

After the injury, blood clots (which take place in the hemostasis stage) block the damaged vessels, and phagocytes such as neutrophils and macrophages enter into the injured areas to clean up the debris and supply GFs to aid in the proliferation stage. To heal the wounds, re-epithelialize, restore vascular systems, and create granulation, many cell types, including keratinocytes, MSCs, endothelial progenitors, and fibroblasts, travel to the site of the injury.

a) **Chitosan** exhibits hemostatic properties through the first healing cascade and encourages neutrophil infiltration/migration, and macrophage, and enhances fibroblast migration as well as collagen deposition, as seen in the case of diabetic mice. They have proper biocompatibility, biodegradability, antibacterial performance, and mucoadhesive feature. It is found in hydrogel, hydrocolloid, film, porous scaffolds, and nanofibrous mats. Chitosan can be manufactured into porous scaffolds or films that give cells a place to connect, grow, and regenerate tissue. Chitosan's porous structure makes it possible for cells to penetrate while also allowing for the stream of nutrients and oxygen. Positive surface charges on chitosan have the effect of promoting cell attachment and facilitating cell growth. It interacts with negatively charged cell membranes, facilitating cell adhesion and improving cell–substrate interaction. By preventing the generation of pro-inflammatory cytokines and encouraging the release of anti-inflammatory substances, chitosan demonstrates anti-inflammatory effects. This characteristic may aid in lowering swelling at the site of the incision and promoting tissue recovery.[19]

b) **Alginate** keeps the wound wet, minimizes bacterial activity, and speeds up the healing process. Enhancing cellular functions are calcium alginates. Alginate hydrogels are commonly employed in skin tissue engineering due to their biocompatibility, biodegradability, capacity for hydrogel formation, and versatility in scaffold construction. Their special qualities make them useful for skin tissue engineering applications that encourage cell development, direct tissue regeneration, and expedite wound healing. Alginate's capacity to create hydrogels in the presence of divalent cations, such as calcium ions, is one of its main benefits. Alginate gels when come in contact with calcium ions, create a three-dimensional network structure. By altering the concentration of calcium ions or by utilizing chelating chemicals that can sequester the calcium ions, one can control the gelation process and direct it to happen at particular times or places. Alginate hydrogels' porous design offers a favorable environment for cell adhesion, proliferation, and migration. To replicate the mechanical characteristics of natural skin tissue, the alginate content, crosslinking density, or addition of other materials can be altered in the hydrogel. Alginate hydrogels can also be made useful by adding bioactive compounds, including peptides or growth factors, to improve cell signaling and encourage tissue regeneration. To direct cellular behavior and encourage tissue regeneration, these bioactive compounds can either be immobilized within the alginate matrix or released under controlled conditions. Additionally, the high water content of alginate hydrogels contributes to the preservation of the moist environment required for cell survival and tissue regeneration.

Additionally, they have strong gas permeability, which promotes the exchange of nutrients and oxygen while eliminating waste.[23]

c) **Gelatin** is a frequently utilized biomaterial in skin tissue engineering due to its favorable characteristics, biocompatibility, and capacity to assist cell proliferation and tissue regeneration. Collagen, a key component of the extracellular matrix (ECM) in many tissues, including skin, is whence gelatin gets its name. It is made by partially hydrolyzing animal tissues that are high in collagen, typically from bovine or porcine sources. To create a supporting matrix for cell attachment, proliferation, and tissue regeneration, gelatin can be processed into a variety of shapes, including films, sponges, hydrogels, or electrospun fibers. The porosity design of gelatin scaffolds makes it easier for cells to infiltrate and diffuse nutrients and oxygen. Gelatin is a superb biomaterial for tissue engineering since it is structurally and biochemically similar to the ECM. It promotes cell proliferation, migration, and differentiation by offering a biochemical and mechanical environment that closely replicates that of native tissue.

d) **Collagen** is an important biomaterial that is widely utilized in skin tissue engineering due to its excellent biocompatibility, biodegradability, and structural resemblance to the skin's original extracellular matrix (ECM). Collagen, the primary protein found in connective tissues, gives the skin its mechanical strength and support, making it a perfect substance for fostering tissue regeneration. The arginine–glycine–aspartate (RGD) sequence is one of the cell-binding domains it has, which helps cells interact with the scaffold. Cells can adhere and disseminate as a result, aiding in the regeneration process. To create a three-dimensional framework that resembles the natural ECM, it can be processed into a variety of shapes, including sponges, films, hydrogels, or fibers. Collagen scaffolds provide an environment that is conducive to cell adhesion, growth, migration, and tissue regeneration. Collagen scaffolds have the ability to promote angiogenesis, which aids in the development of new blood vessels. This is necessary to give the healing tissue nutrition and oxygen. Endothelial cells, which are in charge of blood vessel creation, can migrate and grow more quickly when they are supported by scaffolds made of collagen.[24]

e) **Fibrin,** a protein that aids in blood clotting, is frequently used in skin tissue engineering procedures. Fibrin-based materials provide a number of benefits, including the capacity to encourage cell adhesion, migration, and tissue regeneration. They are also biocompatible and biodegradable. It can be made into films or hydrogels that offer a three-dimensional structure to assist cell adhesion, growth, and tissue regeneration. Fibrin scaffolds can be easily customized to the required shape and size and resemble the natural temporary matrix created during wound healing. Its hydrogels contain adhesion qualities that encourage cell migration and attachment. Fibrin has binding sites for cell surface receptors, which makes it easier for cells to interact with the scaffold. As a result, cells can populate the scaffold and move into the area of the wound to encourage tissue regeneration. Fibrin can be employed to transport bioactive substances like growth factors. These molecules have the ability to be absorbed into the fibrin matrix and then released in a regulated way, supplying localized stimulation to direct cellular behavior, encourage tissue regeneration, and improve wound healing. Wound-healing applications have made use of fibrin-based substances like fibrin sealants and fibrin matrices. To aid in hemostasis and encourage wound closure, fibrin sealants can be applied to wounds. As dermal substitutes, fibrin matrices can stimulate tissue regeneration and speed up wound healing.[25]

### *5.3.4 NEURAL TISSUE ENGINEERING*

#### *5.3.4.1 ROLE OF NATURAL BIOPOLYMER-BASED SCAFFOLDS IN NEURAL TISSUE REGENERATION*

Scaffolds are characterized as three-dimensional porous and stable biomaterials strategically engineered to encompass features such as the promotion of cell–biomaterial interactions, facilitation of cell adhesion, and extracellular matrix (ECM) deposition. These structures also facilitate the exchange of oxygen, nutrients, and signaling factors, exhibit biodegradability, and are noncytotoxic in nature. The original biomaterials employed and still used today had a natural origin and a nature that was identical to or extremely comparable to the tissue that needed to be repaired.[26] The use of natural biomaterials is encouraged since they are compounds that the host cells can understand, resulting in less rejection. Natural polymers generally encourage great cell attachment and development. Not only have matrices composed of generic mixtures of extracellular matrix (ECM) substances been utilized, but also scaffolds constructed from purified ECM molecules such as collagen,[27] hyaluronic acid, and some polysaccharides.[28,29] Natural biomaterials can be produced in a wide range of formats, including in-situ gellable solutions, scaffolds, fibers, and tubular conduits. However, the reality is that these materials lack suitable mechanical characteristics and frequently biodegrade too quickly to provide adequate support.[30]

Scaffolds made of natural biopolymers are essential for the regeneration of brain tissue because they offer a three-dimensional (3D) framework for cell adhesion, proliferation, and differentiation. These scaffolds are often made of biocompatible and biodegradable substances that come from plants, such as proteins, polysaccharides, or mixtures of these substances. Natural biopolymer-based scaffolds play the following important roles in the regeneration of brain tissues. Cell migration and adhesion: Scaffolds based on biopolymers provide adhesive sites that enhance the attachment of neural cells, including neurons, astrocytes, and oligodendrocytes. Cell-adhesive peptides or growth factors can be added to these scaffolds to improve cell adherence and migration into the scaffold, which will speed up tissue regeneration. Delivery of growth factors: Natural biopolymers can be altered to include growth factors or neurotrophic factors, which are crucial for the regeneration of neural tissue. The scaffolds can provide continuous and localized distribution by managing the kinetics of these factors' release, which encourages cell division and proliferation while directing axonal outgrowth. Structural support and guidance: The regenerated neural tissue is mechanically supported by the biopolymer-based scaffolds' 3D structure. They can be designed to have a particular porosity, pore size, and interconnectivity, which permits cell infiltration and makes it easier to exchange nutrients and waste. Additionally, these scaffolds can be made to incorporate topographically aligned fiber structures, which can have biodegradability: Natural biopolymers are frequently biodegradable, which means that over time, natural processes can break them down. Because the scaffolds can progressively deteriorate and be replaced by new tissue without requiring additional surgical intervention, this characteristic is advantageous in brain tissue regeneration. Biopolymer breakdown byproducts can typically be metabolized or eliminated from the body without harm.

Immunomodulation: Some organic biopolymers are naturally immunomodulatory. At the location of the damage or implantation, they can modify the immune response, lowering inflammation and promoting a healing environment. This immunomodulatory impact can reduce the production of brain tissue, and direct the regeneration of brain tissue direct the regenerating brain cells and encourage axonal. Overall, natural biopolymer-based scaffolds offer a biomimetic and supportive environment for neural tissue regeneration. Their ability to promote cell adhesion, guide cell behavior, deliver growth factors, and gradually degrade makes them valuable tools in tissue engineering strategies aimed at repairing and regenerating damaged neural tissues.

### 5.3.5 CARDIOVASCULAR TISSUE ENGINEERING

#### 5.3.5.1 NATURAL BIOPOLYMERS FOR CARDIAC AND VASCULAR TISSUE REGENERATION

The adult mammalian heart is one of the least regenerable organs; it is vulnerable to numerous threats, including necrosis, apoptosis, and oncosis (or ischemia cell death), which can result in heart failure.[31,32] Recently, it has been discovered that apoptosis, or programmed cell death, and necrosis, or premature cell death brought on by physical or chemical insult, coexist during pathological states of cardiac disease.[33] Concerning cardiac pathogenesis, myocardial infarction leads to the formation of scar tissue, characterized by areas where cardiomyocytes are substituted with fibrillar collagen, and or probably cells resembling fibroblasts.[34] Necrosis and oncosis both occur after cell injury,[35] but oncosis, also known as ischemic cell death, is distinguished from necrosis by the fact that the cell swells rather than contracts. As of 2017, 38 million individuals worldwide,[36] with the estimates in the United States being 6.5 million people,[37] were affected by heart failure. Along with cardiac pathophysiology, age-related increases in the risk of heart disease are both gradual and abrupt.[38] All of these elements compete with the mature mammalian CMs' modest annual cell turnover rates, which range from 0.3 to 1%. The heart is the most noteworthy arena of study for tissue engineering for all of these reasons and more. These studies would not provide only new scientific insights into the mechanisms of heart repair and improvement through tissue engineering, Moreover, these discoveries could propel advancements in developing novel therapeutic approaches for clinical interventions.

The regeneration of cardiac and vascular tissue depends heavily on natural biopolymers. These biopolymers have several benefits, including the capacity to stimulate cellular growth and tissue regeneration as well as biocompatibility and biodegradability. Here are some instances of typical natural biopolymers for regenerating heart and vascular tissue:

- **Collagen:** The most prevalent protein in the human body is collagen, which is also a crucial part of the extracellular matrix (ECM) in cardiac and vascular tissues. It encourages cell attachment and migration while also offering structural support. It is possible to create a supportive environment for cell development and tissue regeneration using hydrogels or scaffolds made of collagen.
- **Fibrin:** An essential component of blood clotting is fibrin, a fibrous protein. It has been widely employed as a natural biopolymer for cardiac and vascular tissue engineering and can be generated from human or animal blood. Fibrin scaffolds support vascularization, cell proliferation, and attachment.
- **Chitosan:** Chitin, a natural polysaccharide present in the exoskeletons of crustaceans, is the source of chitosan, a substance that is produced from it. Additionally, it is biocompatible and has antimicrobial qualities. Scaffolds made of chitosan can be utilized in vascular and cardiac applications to assist cell adhesion, migration, and tissue regeneration.
- **Gelatine:** Gelatine is produced from collagen and shares characteristics with collagen, such as promoting cell attachment and being biocompatible. To facilitate tissue regeneration in cardiac and vascular applications, it is easily processed into a variety of forms, such as hydrogels and sponges.
- **Hyaluronic acid (HA):** Found in connective tissues, hyaluronic acid is a naturally occurring polysaccharide. It helps tissues stay hydrated and lubricated and is crucial for cell signaling and wound healing. Hydrogels or scaffolds made of HA can be utilized to encourage cell migration, growth, and tissue regeneration.
- **Silk fibroin:** Derived from the cocoon of the silkworm, silk fibroin is highly biocompatible and has great mechanical qualities. It can

be processed into a variety of shapes, including films, scaffolds, and fibers, to aid in the regeneration of heart and vascular tissue.

To make scaffolds or hydrogels that mimic the natural extracellular matrix and offer a favorable microenvironment for cell attachment, proliferation, and tissue regeneration in cardiac and vascular applications, these natural biopolymers can be utilized individually or in conjunction with synthetic materials or growth factors.

## 5.4 REGULATORY CONSIDERATIONS FOR NATURAL BIOPOLYMER-BASED SCAFFOLDS

Regulatory organizations like the European Medicines Agency (EMA) and the U.S. Food and Drug Administration (FDA) need thorough safety and effectiveness evidence for biopolymer-based scaffolds. This involves proof of the scaffold's biocompatibility, lack of toxicity, and effective operation in both preclinical and clinical investigations. Biocompatibility testing: To assess how the scaffold interacts with the surrounding tissues or cells, regulatory authorities may impose thorough biocompatibility testing, including in-vitro and in-vivo research. These studies evaluate the scaffold's breakdown products, immunogenicity, inflammation, and possible adverse effects. Manufacturing under Good Manufacturing Practices (GMP) is often required by regulations to guarantee scaffolds are produced with uniform quality, sterility, and repeatability. Validating appropriate sterilization techniques is necessary to make sure that the scaffold is free of pathogens and meets regulatory requirements. Preclinical and clinical studies may be required to prove the safety and efficacy of the biopolymer-based scaffold, depending on the regulatory authority. These investigations must comply with strict regulatory requirements and have a thorough design.

## 5.5 CONCLUSION

For the ideal design of a biomaterial scaffold system for effective tissue replacement therapies, a better understanding of the native tissue structure, physical and biochemical properties are some essential factors to be considered. When a damage or defect in a biological tissue cannot be effectively restored by the body's natural recovering abilities, in order to stimulate functional healing, biopolymer-based scaffolds can be utilized as part of a therapeutic tissue engineering strategy. Biopolymer scaffolds play immense potential as candidate materials for this task, injury-specific and tissue-specific applications are a subject of immense research which needs fine-tuning of their properties. Here, in this chapter, we give an outline of some of the most pronounced materials and important methods employed for producing biopolymer tissue scaffolds and highlight research using biopolymer scaffolds for regenerative medicine.

## KEYWORDS


- **regenerative medicine**
- **scaffolds**
- **natural biopolymers**
- **clinical applications**
- **tissue engineering**
- **biomaterials**
- **biocompatibility**
- **biodegradability**


## REFERENCES


1. Vimalraj, S.; Renugaa, S.; Dhanasekaran, A. Chick Embryo Chorioallantoic Membrane: A Biomaterial Testing Platform for Tissue Engineering Applications. *Process Biochem*. **2023,** *124*, 81–91. https://doi.org/10.1016/J.PROCBIO.2022.11.007.
2. Reddy, M.S.B.; Ponnamma, D.; Choudhary, R.; Sadasivuni, K. K. A Comparative Review of Natural and Synthetic Biopolymer Composite Scaffolds.; *Polymers* **2021,** *13*, 1105. https://doi.org/10.3390/POLYM13071105.
3. George, A. M.; Reddy Peddireddy, S. P.; Thakur, G.; Rodrigues, F. C. Biopolymer-Based Scaffolds: Development and Biomedical Applications. *Biopolym.-Based Formulations: Biomed. Food App*. **2020,** 717–749. https://doi.org/10.1016/B978-0-12-816897-4.00029-1.

4. Chen, F. M.; Liu, X. Advancing Biomaterials of Human Origin for Tissue Engineering. *Prog. Polym. Sci.* **2016,** *53*, 86–168. https://doi.org/10.1016/j.progpolymsci.2015.02.004.
5. Babaie, E.; Bhaduri, S. B. Fabrication Aspects of Porous Biomaterials in Orthopedic Applications: A Review.; *ACS Biomater. Sci. Eng.* **2018,** *4*, 1–39. https://doi.org/10.1021/ACSBIOMATERIALS.7B00615/ASSET/IMAGES/MEDIUM/AB-2017-00615V_0017.GIF.
6. Coburn, J.; Gibson, M.; Bandalini, P. A.; Laird, C.; Mao, H. Q.; Moroni, L.; Seliktar, D.; Elisseeff, J. Biomimetics of the Extracellular Matrix: An Integrated Three-Dimensional Fiber-Hydrogel Composite for Cartilage Tissue Engineering. *Smart Struct. Syst.* **2011,** *7*, 213. https://doi.org/10.12989/SSS.2011.7.3.213.
7. Chan, B. P.; Leong, K. W. Scaffolding in Tissue Engineering: General Approaches and Tissue-Specific Considerations. *Eur. Spine J.* **2008,** *17*, 467–479. https://doi.org/10.1007/S00586-008-0745-3.
8. Hollister, S. J. Porous Scaffold Design for Tissue Engineering. *Nat. Mater.* **2005,** *4*, 518–524. https://doi.org/10.1038/nmat1421.
9. Tripathy, N.; Perumal, E.; Ahmad, R.; Song, J. E.; Khang, G. Hybrid Composite Biomaterials. *Principles Regen. Med.* (**2019,**695–714. https://doi.org/10.1016/B978-0-12-809880-6.00040-0.
10. Kim, S.; Uroz, M.; Bays, J. L.; Chen, C. S. Harnessing Mechanobiology for Tissue Engineering.; *Dev Cell* **2021,** 56, 180–191. https://doi.org/10.1016/J.DEVCEL.2020.12.017.
11. Liu, S.; Yu, J.-M.; Gan, Y.-C.; Qiu, X.-Z.; Gao, Z.-C.; Wang, H.; Chen, S.-X.; Xiong, Y.; Liu, G.-H.; Lin, S.-E.; McCarthy, A.; John, J. V.; Wei, D.-X.; Hou, H.-H. Biomimetic Natural Biomaterials for Tissue Engineering and Regenerative Medicine: New Biosynthesis Methods, Recent Advances, and Emerging Applications. *Mil. Med. Res.* **2023,** *10*, 16. https://doi.org/10.1186/S40779-023-00448-W.
12. Troy, E.; Tilbury, M. A.; Power, A. M.; Wall, J. G. Nature-Based Biomaterials and Their Application in Biomedicine.; *Polymers* **2021,** *13*. https://doi.org/10.3390/POLYM13193321.
13. Pramanik, S.; Kharche, S.; More, N.; Ranglani, D.; Kapusetti, G. Natural Biopolymers for Bone Tissue Engineering: A Brief Review. *Eng. Regen.* **2023,** *4*, 193–204. https://doi.org/10.1016/J.ENGREG.2022.12.002.
14. Turnbull, G.; Clarke, J.; Picard, F.; Riches, P.; Jia, L.; Han, F.; Li, B.; Shu, W. 3D Bioactive Composite Scaffolds for Bone Tissue Engineering. *Bioact. Mater*. **2018,** *3*, 278–314. https://doi.org/10.1016/J.BIOACTMAT.2017.10.001.
15. Polo-Corrales, L.; Latorre-Esteves, M.; Ramirez-Vick, J. E. Scaffold Design for Bone Regeneration. *J. Nanosci. Nanotechnol.* **2014,** *14*, 15. https://doi.org/10.1166/JNN.2014.9127.
16. Stamnitz, S.; Klimczak, A. Mesenchymal Stem Cells, Bioactive Factors, and Scaffolds in Bone Repair: From Research Perspectives to Clinical Practice. *Cells* **2021,** *10*. https://doi.org/10.3390/CELLS10081925.
17. Qiang Ruan, S.; Deng, J.; Yan, L.; Liang Huang, W. Composite Scaffolds Loaded with Bone Mesenchymal Stem Cells Promote the Repair of Radial Bone Defects in Rabbit Model. *Biomed. Pharmacother*. **2018,** *97*, 600–606. https://doi.org/10.1016/J.BIOPHA.2017.10.110.

18. Wei, W.; Ma, Y.; Yao, X.; Zhou, W.; Wang, X.; Li, C.; Lin, J.; He, Q.; Leptihn, S.; Ouyang, H. Advanced Hydrogels for the Repair of Cartilage Defects and Regeneration. *Bioact. Mater.* **2021,** *6* (4), 998–1011.
19. Ahmadi, F.; Oveisi, Z.; Mohammadi Samani, S.; Amoozgar, Z. Chitosan Based Hydrogels: Characteristics and Pharmaceutical Applications. *Res. Pharm. Sci.* **2015,** *10* (1), 1.
20. Szustak, M.; Gendaszewska-Darmach, E. Nanocellulose-Based Scaffolds for Chondrogenic Differentiation and Expansion. *Front. Bioeng. Biotechnol.* **2021,** *9*, 736213.
21. Nguyen, T. P.; Nguyen, Q. V.; Nguyen, V. H.; Le, T. H.; Huynh, V. Q. N.; Vo, D. V. N.; Trinh, Q. T.; Kim, S. Y.; Van Le, Q. Silk Fibroin-Based Biomaterials for Biomedical Applications: A Review. *Polymers* **2019,** *11* (12), 1933.
22. Irawan, V.; Sung, T.-C.; Higuchi, A.; & Ikoma, T. (2018). Collagen Scaffolds in Cartilage Tissue Engineering and Relevant Approaches for Future Development. *Tissue Eng. Regen. Med.* **2018,** *15* (6), 673. https://doi.org/10.1007/S13770-018-0135-9
23. Abasalizadeh, F.; Moghaddam, S. V.; Alizadeh, E.; Akbari, E.; Kashani, E.; Fazljou, E. M. B.; Torbati, M.; Akbarzadeh, A. Alginate-Based Hydrogels as Drug Delivery Vehicles in Cancer Treatment and Their Applications in Wound Dressing and 3D Bioprinting. *J. Biol. Eng.* **2020,** *14*, 1–22.
24. Zarrintaj, P.; Seidi, F.; Youssefi Azarfam, M.; Khodadadi Yazdi, M.; Erfani, A.; Barani, M.; Chauhan, N. P. S.; Rabiee, N.; Kuang, T.; Kucinska-Lipka, J.; Saeb, M. R.; Mozafari, M. Biopolymer-Based Composites for Tissue Engineering Applications: A Basis for Future Opportunities. *Compos. Part B: Eng.* **2023,** *258*, 110701. https://doi.org/10.1016/j.compositesb.2023.110701
25. Schultz, G.; Chin, G.; Moldawer, L.; Diegelmann, R. *Ch23 Mechanisms of Vascular Disease*; University of Adelaide Press, 2011.
26. Doblado, L. R.; Martínez-Ramos, C.; Pradas, M. M. Biomaterials for Neural Tissue Engineering. *Front. Nanotechnol.* **2021,** *3*, 21. https://doi.org/10.3389/FNANO.2021.643507
27. Bozkurt, A.; Brook, G. A.; Moellers, S.; Lassner, F.; Sellhaus, B.; Weis, J.; Woeltje, M.; Tank, J.; Beckmann, C.; Fuchs, P.; Damink, L. O.; Schügner, F.; Heschel, I.; Pallua, N. In Vitro Assessment of Axonal Growth Using Dorsal Root Ganglia Explants in a Novel Three-Dimensional Collagen Matrix. *Tissue Eng.* **2007,** *13*, 2971–2979. https://doi.org/10.1089/TEN.2007.0116
28. Austin, J. W.; Kang, C. E.; Baumann, M. D.; DiDiodato, L.; Satkunendrarajah, K.; Wilson, J. R.; Stanisz, G. J.; Shoichet, M. S.; Fehlings, M. G. The Effects of Intrathecal Injection of a Hyaluronan-Based Hydrogel on Inflammation, Scarring and Neurobehavioural Outcomes in a Rat Model of Severe Spinal Cord Injury Associated with Arachnoiditis. *Biomaterials* **2012,** *33*, 4555–4564. https://doi.org/10.1016/J.BIOMATERIALS.2012.03.022
29. Mammadov, B.; Mammadov, R.; Guler, M. O.; Tekinay, A. B. Cooperative Effect of Heparan Sulfate and Laminin Mimetic Peptide Nanofibers on the Promotion of Neurite Outgrowth. *Acta Biomater.* **2012,** *8*, 2077–2086. https://doi.org/10.1016/J.ACTBIO.2012.02.006
30. Terenghi, G. Peripheral Nerve Regeneration and Neurotrophic Factors. *J Anat.* **1999,** *194* (Pt 1), 1–14. https://doi.org/10.1046/J.1469-7580.1999.19410001.

31. Heallen, T. R.; Martin, J. F. Heart Repair via Cardiomyocyte-Secreted Vesicles. *Nat. Biomed. Eng.* **2018,** *2*, 271–272. https://doi.org/10.1038/S41551-018-0239-5.
32. Mohamed, T. M. A.; Ang, Y. S.; Radzinsky, E.; Zhou, P.; Huang, Y.; Elfenbein, A.; Foley, A.; Magnitsky, S.; Srivastava, D. Regulation of Cell Cycle to Stimulate Adult Cardiomyocyte Proliferation and Cardiac Regeneration. *Cell* **2018,** *173*, 104–116. e12. https://doi.org/10.1016/J.CELL.2018.02.014.
33. Frangogiannis, N. G. The Functional Pluralism of Fibroblasts in the Infarcted Myocardium. *Circ. Res.* **2016,** *119*, 1049. https://doi.org/10.1161/CIRCRESAHA. 116.309926.
34. Weerasinghe, P.; Buja, L. M. Oncosis: An Important Non-Apoptotic Mode of Cell Death. *Exp. Mol. Pathol.* **2012,** *93*, 302–308. https://doi.org/10.1016/J.YEXMP.2012. 09.018.
35. Tzahor, E.; Poss, K. D. Cardiac Regeneration Strategies: Staying Young at Heart. *Science* **2017,** *356*, 1035–1039. https://doi.org/10.1126/SCIENCE.AAM5894/ ASSET/612504F3-BF6C-419F-AE43-EA07748C57B5/ASSETS/GRAPHIC/ 356_1035_F4.JPEG.
36. Benjamin, E. J.; Blaha, M. J.; Chiuve, S. E.; Cushman, M.; Das, S. R.; Deo, R.; De Ferranti, S. D.; Floyd, J.; Fornage, M.; Gillespie, C.; Isasi, C. R.; Jimenez, M. C.; Jordan, L. C.; Judd, S. E.; Lackland, D.; Lichtman, J. H.; Lisabeth, L.; Liu, S.; Longenecker, C. T.; MacKey, R. H.; Matsushita, K.; Mozaffarian, D.; Mussolino, M. E.; Nasir, K.; Neumar, R. W.; Palaniappan, L.; Pandey, D. K.; Thiagarajan, R. R.; Reeves, M. J.; Ritchey, M.; Rodriguez, C. J.; Roth, G. A.; Rosamond, W. D.; Sasson, C.; Towfghi, A.; Tsao, C. W.; Turner, M. B.; Virani, S. S.; Voeks, J. H.; Willey, J. Z.; Wilkins, J. T.; Wu, J. H. Y.; Alger, H. M.; Wong, S. S.; Muntner, P. Heart Disease, and Stroke Statistics-2017 Update: A Report From the American Heart Association. *Circulation* **2017,** *135*, e146–e603. https://doi.org/10.1161/CIR.0000000000000485.
37. Dhingra, R.; Vasan, R. S. Age as a Risk Factor. *Med. Clin. North Am.* **2012,** *96*, 87–91. https://doi.org/10.1016/J.MCNA.2011.11.003.
38. Zhao, Y.; Eng, G.; Lee, B. W.; Radisic, M.; Vunjak-Novakovic, G. Cardiac Tissue Engineering. *Principles Tissue Eng.* **2020,** 593–616. https://doi.org/10.1016/B978-0-12-818422-6.00033-2.
39. Biswas, M. C.; Jony, B.; Nandy, P. K.; Chowdhury, R. A.; Halder, S.; Kumar, D.; Ramakrishna, S.; Hassan, M.; Ahsan, M. A.; Hoque, M. E.; Imam, M. A. Recent Advancement of Biopolymers and Their Potential Biomedical Applications. *J. Polym. Environ.* **2021,** 1–24.

# CHAPTER 6

# Sustainable and Innovative Food Packing Materials Based in Agroindustrial By-Products

SANDY R. GARCÍA-RODRÍGUEZ[1], ALAIN MARTÍNEZ-PÉREZ[2], CHRISTIAN I. CANO-GÓMEZ[1], DIANA B. MUÑIZ-MÁRQUEZ[1], and JORGE E. WONG PAZ[1]

[1]*Facultad de Estudios Profesionales Zona Huasteca, Universidad Autónoma de San Luis Potosí, Ciudad Valles, San Luis Potosí, México*

[2]*Departamento de Ingeniería Química y Bioquímica, TecNM/Instituto Tecnológico de Durango, Durango, Durango, México*

## ABSTRACT

The utilization of agroindustrial by-products and alternative materials in the evolution of novel food packaging represents a sustainable approach, given the substantial environmental problems associated with certain conventional materials. Concurrently, contemporary food industries necessitate the evolution of innovative packaging designs to align with consumer preferences. In this context, the incorporation of sustainable packaging materials, such as biopolymers derived from plant-based sources, proves instrumental in the creation of edible films, bioadhesives, coatings, plastic films, and various other emerging forms of food packaging. This strategic integration aims to mitigate the accumulation of food wastes generated across diverse industries. This chapter elucidates the pivotal role played by innovative food packaging in addressing and

Polymers and Functional Materials for a Cleaner Environment and Human Health: Sustainable Technologies and New Research. Ruby Varghese, Ann Rose Abraham, Shrikaant Kulkarni, & A. K. Haghi (Eds.)

ameliorating sustainable challenges. Furthermore, it delves into recent scientific advancements in this subject, highlighting the ongoing efforts to foster sustainable practices within the food packaging industry.

## 6.1 INTRODUCTION

The implementation of food packaging arose from the need of people to preserve their food for longer periods of time; thus, it was adapted to industrial and commercial processes.[1] Currently, the food industry presents a great challenge, to meet the demands of consumers who have a tendency toward fresh, healthy food and with a lower percentage of processing; this related to the need for better options to improve the style of their lives in terms of less time to do their shopping at the supermarket and therefore spend less time cooking food. Based on this, it is known that consumers are aware of the great importance of the impact of food packaging for its conservation and likewise with the environment because of the origin of its compounds.[2,3] The globalization of many products has made them more accessible worldwide, so that people can purchase them at any time and at a very low price.[4] Globally, the development of food packaging is one of the industries in which a lot of technology and time is invested, which is why it has become one of the most advanced, which shows the great impact on the economy of this industrial sector.[5]

The significance of product packaging in the distribution chain cannot be overstated, encompassing transport, delivery, and retail stages. Its pivotal role extends to ensuring the protection and preservation of food items. However, a concerning observation underscores the impact of inadequate packaging on food waste, with over 25% of food products reportedly being discarded due to various packaging-related issues. These issues include imprecise dosing, the absence of replenishment functionality, insufficient storage protection, and other factors.[6] Solving these problems related to food waste, food packaging technologies are still advanced, so the producers of the packaging market are developing new models of packaging with trends such as packaging that are environmentally friendly and have a benefit to the consumer; in the same way, they are ready to eat and with a longer shelf life thanks to active and intelligent packaging know-how.[7] The rationale of food packaging is to achieve shelf life extension by making a barrier as a protection from the outside environment.[8] Thanks to new technologies it is easier to implement strategies to help extend the shelf life of a product such as chemical additions, oxygen indicators, nanomaterials, biosensors, biofilms, bioadhesives, among others.[9–11]

The cost of packaging contributes significantly to the overall cost of packaged foods, and consumers are known to be aware of this factor. Over the past few years, a growing emphasis on environmental sustainability, healthier dietary choices, and addressing challenges related to food safety, storage, and transportation has underscored the demand for biologically based plastics and sensors. This shift is driven by a desire to find alternatives derived from agro-industrial products, which can alleviate the environmental impact and mitigate public health and economic concerns associated with traditional materials.[12]

## 6.2 IMPORTANCE OF FOOD PACKING

Food packaging is an essential element within the food industry, impacting various aspects of food, from safety to nutritional quality.[13] Not only does packaging play a key role in preserving and protecting food, but it also acts as an important communication tool between producers and consumers. The appropriate choice of packaging can influence the lifespan and quality of a food product, as well as its environmental impact and economic expense.

Packaging affects several areas: it determines the acceptability of the product, nutritional quality, and, most importantly, food safety. Additionally, it provides consumers with an initial impression of the product, influencing their buying decisions. In today's consumer landscape, there is a growing demand for innovative packaging solutions, including active and intelligent packaging. These advanced systems interact with the food environment, either releasing certain substances or absorbing others, ultimately prolonging the life of food products.[14]

Nevertheless, achieving a right balance in packaging is crucial. Insufficient packaging can jeopardize product safety, while excessive packaging leaves a significant environmental footprint due to waste. Therefore, it is important to balance energy costs, material expenses, environmental impacts, contamination risks, and food protection.[13]

Modern packaging for food products is diverse and complex. There are three primary levels: primary, secondary, and tertiary packaging. Primary packaging directly encases food and is in direct contact with it, greatly influencing food quality. Secondary packaging holds products enclosed within the primary packaging, while tertiary packaging is the external layer used for bulk handling, distribution, and storage. Each level has its own importance, but even with these three layers, food safety is not always assured. Packaging serves multiple purposes: it facilitates transportation, maintains food quality, protects the integrity of the food, shields food from harmful elements and chemicals, and prevents the proliferation of bacteria and pests. Additionally, packaging provides consumers with vital product information through labeling, such as product names, ingredient lists, consumption methods, pricing, and expiry dates.[15]

The advancement of food packaging in recent years has been noteworthy, reflecting society's increasing awareness about maintaining food quality. Packaging not only shields its contents from deterioration caused by various organisms but also ensures food quality, safety, and extends shelf life. Historically, materials like paper, cardboard, metal, glass, and plastic were utilized. Factors such as the global food market growth, government regulations, availability of raw materials, and consumer tastes have continually shaped the food packaging sector.[16] Materials used for packaging must provide mechanical, physical, and thermal robustness, in addition to optical, protective, and antimicrobial features. In today's world, packaging is almost omnipresent, and most consumer goods on the market are seen in a limited variety of packaging materials.[17,18]

However, interactions between food and its packaging can sometimes adversely affect quality and safety. For example, when food is packed in polymer-based materials, there is potential for flavor alterations due to scent absorption and the migration of undesirable flavors from the packaging. Thus, selecting suitable packaging materials and creating active, intelligent packaging depends on the packaging's intentional interaction with food. The primary aim is to improve food quality and safety while maximizing product longevity and minimizing undesired changes.[19]

Yet, there is an environmental concern. Most current packaging materials are nondegradable and pose environmental challenges. A potential solution is edible and biodegradable films, which can reduce packaging waste while improving the lifespan and quality of food products. However, the use of bio-based materials for food packaging is restricted, mainly due to their weaker barrier and structural properties.[20]

As society advances and consumer preferences evolve, the food industry must continually adapt, ensuring that packaging meets safety standards, environmental concerns, and consumer needs. As highlighted by multiple studies,[13–16] the significance of food packaging in our contemporary society is undeniable, serving both functional and communicative roles.

## 6.3 ACTIVE FOOD PACKING SYSTEMS

Plastic films, due to their versatility, low cost, and low weight, are one of the most widely used materials for food packaging, as they act as a barrier to food spoilage causes such as humidity, temperature, soil, and microorganisms, among others.[21] Recent advances in food packaging incorporate substances that provide additional protection to ordinary packaging due to the active components incorporated into the packaging matrix. This new type of packaging is known as active packaging.[22]

Petroleum-derived polymers are the material for excellence for food packaging thanks to their physicochemical properties and easy production. However, current trends revolve around green technology, circular economy, and environmental care, so naturally occurring polymers such as chitosan, cellulose, starch, agar, and gelatin have some advantages.[22] An alternative to traditional petroleum-derived plastic packaging is hydrogels, which are chains of natural or synthetic polymers cross-linked three-dimensionally and can retain large volumes of water maintaining their integral structure. Despite their great affinity for water, they do not dissolve and allow the integration of bioactive compounds that improve functionality and physicochemical properties.[23] For their part, nanomaterials have physicochemical properties greater than some conventional materials, so the research and application of these materials in the packaging design for the food and pharmaceutical industry is increasing.[21]

For a package to be considered active, it must incorporate bioactive agents that are naturally released inside or outside the food to protect of food spoilage factors such as light, microorganisms, moisture, and oxygen, among others. The main active compounds incorporated into active packaging include antioxidants, antimicrobials, oxygen scavengers, and ethylene scavengers, which provide excellent potential for preserving the quality and safety of food products, particularly highly perishable foods.[22,24]

The main innovations in the technological development of active packing focus on integrating substances capable of inhibiting or preventing microbial growth in foods. Some antimicrobial substances used are plant-based organosulfur compounds, essential oils, chitosan, polyphenols, and silver-based additives.[25] Often, substances used as antioxidants fulfill other bioactivities such as antimicrobials. There are two types of antioxidant substances used in active packing of synthetic origin, such as butylated hydroxyanisole (BHA) and butylated hydroxytoluene (BHT), and natural ones, such as polyphenols, essential oils, and phytic acid.[26] These substances are added to a natural or synthetic polymer-based coating, which must be supplied in a controlled manner, neither too low so as not to exert its activity nor too high so as to alter the sensory characteristics of the food or to have toxicological effects on the consumer.[25] Nanocomposites are a contribution of nanotechnology to advanced food packaging technology, particularly when it comes to nanocomposites with adjustable properties such as heat and moisture resistance. Zinc oxide, titanium dioxide, and silver nanoparticles are the best representatives. These nanomaterials provide mechanical resistance and resistance against air diffusion and flavors.[27] Ethylene is a phytohormone that causes postharvest deterioration of fruits and vegetables, so managing this hormone's presence is highly important to preserve these foods. Ethylene scavengers such as zeolite, $TiO_2$, and transition metals are used to control the activity of this hormone. Despite numerous studies on the effectiveness of these ethylene scavengers, their application has been limited to depositing these substances inside small sachets and not their integration into the packaging material.[28]

## 6.4 SMART FOOD PACKING SYSTEMS

Different active packaging systems, and smart packaging systems influence and control the physicochemical conditions of the product such as the shelf life of the food, as well as the influence of external factors such as pH, temperature, and the presence of gases such as oxygen and ethylene, during the storage and supply chain. The way to detect these changes that arise inside the packaging is thanks to a device that is designed to detect internal and external changes to provide information to the consumer visually or sonorously about the quality of the product.[22]

### *6.4.1 SENSORS*

A problem with some packaging materials such as cardboard is that they prevent sensory exposure and consumers must rely on the expiration date printed on the package.[29] Sensors used in smart packaging represent one of the great technological advances in food packaging that monitor and control product quality. They are designed so that they can be easily integrated into the packaging, measure freshness and quality markers, and provide a real-time report of food quality indexes such as changes in temperature and the presence of toxic substances that represent a hazard to consumption. These sensors are specially designed to detect a particular analyte by means of chemical or biological receptors, and a physical transducer transforms the response into a qualitative or quantitative measurable signal. The use of sensors in smart packaging for food freshness assessment has been successful, mainly with chemical rather than biological sensors.[30] Thanks to engineering, nanomaterials capable of detecting the presence of gases, aromas, chemical contaminants, and pathogens can be designed, being especially useful for the consumer as the product is selected at its optimum point of freshness and quality.[29]

### *6.4.2 INDICATORS*

Smart indicators are substances that provide information on food quality parameters such as temperature, pH, microbiological status, $CO_2$ concentration, freshness, and physical appearance, and can also detect the presence of leaks.[22] Smart materials can be internal or external depending on the location of the indicator, for example, gas sensors and microbial indicators are usually placed inside the package unlike barcodes and radio frequency identification tags that are placed on the outside of the package.[31] These indicators can be of two types: food freshness and time-temperature indicators. Food freshness indicators were developed to assess microbial growth and chemical changes during food spoilage caused by changes in the concentration of organic acids, glucose, $CO_2$, ethanol, volatile nitrogenous compounds, biogenic amines, and sulphury compounds that usually cause changes in pH resulting in a color variation of the indicator.[32] pH indicators are used in innovative packaging to monitor real-time changes in food quality; some are based on natural pigments and polymers and have great potential application in smart packaging as pH indicators, and edible indicator films can also be developed.[33] On the other hand, time-temperature indicators are devices placed on the outside of the packaging and provide information on the quality of the food along the supply chain.[31] According to their operation mechanism, they can be classified into three groups: physical, chemical, and biological. These indicators can be microbiological, enzymatic, diffusion, solid-state reaction, and polymeric, which are commercially available.[24,34]

### 6.4.3 RADIOFREQUENCY IDENTIFICATION

Radiofrequency identification is carried out by means of tags (data carriers) placed on the outside of the packaging that through wireless communication provide real-time information on product quality status, such as temperature, shelf life, relative humidity, and nutritional information through the supply chain.[22] Radio frequency identification tags are basically composed of three parts, a microchip attached to a small antenna, a reader that receives and emits signals, and a network system or web server.[32] The most well-known sensors in this system are the electronic nose, the electric tongue, and the food sniffer. Although these sensors can accurately measure different quality parameters in food, they have some drawbacks for use in food packaging systems due to their large size, rigidity, and high cost.[35]

## 6.5 BIO-BASED SENSOR DEVELOPMENT BY 3D

Biosensors take advantage of the recognition capacity of biomolecules such as enzymes, aptamers, and antibodies with the aim of improving the selectivity of the analyte, and detection is carried out by means of colorimetry, electrochemistry, optics, and mass. These have made an impact in medicine such as the electrochemical biosensor used for glucose detection; however, in the food industry research and development advances are still limited.[30] Biosensors basically consist of three components, the first is the base material of the sensor which can be glass, metal, paper, or polymer. The second is a bioreceptor, which can be antibodies, enzymes, and cellular structures. Finally, we have the transducer that is responsible for providing a measurable signal obtained by physicochemical changes caused by the bioreceptor when interacting with the analyte; this response must be proportional to the concentration of the interaction of the bioreceptor with the target.[36] The use of biosensors in smart packaging is currently a challenge; most of them are in the early stages of research. Despite the significant advances that have been made in the development of biosensors for the detection of allergens, pathogens, pollutants, generation of toxic substances, and more food quality parameters, reducing the time of analysis of samples with conventional methods, they still have some limitations for their application in smart packaging. One of these problems is the size of the sensors, as they are usually too large to be incorporated into the packaging; another inconvenience is the high cost as they have not yet reached mass development; the food–sensor interaction is also something that must be evaluated in detail as there may be particle migration and these must comply with food production regulations.[30]

Some biosensors as mentioned above are based on natural compounds with the capacity to change color according to pH or the presence of a particular substance or environment. Curcumin is a compound that has been successfully used in smart films as an indicator of the presence of ammonia since this compound is susceptible to acid–base reactions producing changes in coloration. The color change responses in the film based on tara gum, polyvinyl alcohol, and curcumin were between one and three minutes. Furthermore, this film was tested on the spoilage of shrimp, and its color change in the film was evident, demonstrating its potential as a biosensor in the food industry.[37] Maftoonazad and Ramaswamy[38] developed electrospun fiber mats based on polyvinyl alcohol and the natural pigment of red cabbage (*Brassica oleracea L.*) that function as a biosensor indicating a change in pH in a range of 2–12. Furthermore, tests performed with fresh date rutab showed a correlation of pH changes with color changes.

One approach being taken in biosensor development is molecularly imprinted polymers as an alternative to antibodies, allowing functional compounds to bind around the target or analyte.[36] In addition, reporter groups can be integrated into the polymer to increase the sensitivity of the sensor.[39] Printing in 3D is a technology based on three-dimensional mathematical models for rapid and accurate object formation. This technology is being implemented in the food industry in food packaging using extrusion 3D printing for the design of sensors, indicators, and labels.[40] In addition, it allows the use of materials that are less expensive than conventional ones, biocompatible and nontoxic, favoring a more significant presence of smart packaging, reducing losses due to food spoilage, and reducing infections due to consumption of spoiled food.[41] Zhou et al.[42] designed an aerogel based on carboxymethyl nanocellulose, glycerin, and acrylamide derivatives as a shell using coaxial 3D printing technology, while chitosan with silver nanoparticles was immobilized inside, thus obtaining an aerogel with antibacterial and buffering function. The aerogel exhibited good rheological properties such as buffering and antibacterial properties against *E. coli* and *S. aureus* and was also biodegradable. Xue Wang et al. employed 3D printing technology to fabricate a biosensor based on a copper metal-organic framework and tyrosinase enzyme for rapid detection of bisphenol A with high sensitivity and fast detection. In addition, 3D printing can be used for the fabrication of packaging material using bio-based waste materials such as wood flour, grain husk, and fibers. The mechanical properties can be modified by adjusting the powder and binder material concentrations to achieve similar properties to conventionally used polymers.[43]

## 6.6 BIO-BASED FOOD PACKING

In recent years, the food industry has given great importance to a concern about the great implementation of packaging material that is suitable as degradable coatings or biofilms that have thermal, mechanical, and barrier particularities to help the food not to present contamination or its loss and therefore have a significant circular bioeconomy; it is important to know that if the conditions of the food are not taken into account it can be a major problem environmentally and economically.[44] There are strategies for product protection; one of them is that bio-based food grade packaging can protect the food content from external factors, preserve its quality, and improve shelf life. In this way, bio-based raw materials have the potential to be implemented as environmentally friendly materials as substitutes for nonrenewable and nonbiodegradable materials. The analysis of some polymers that are recyclable and biodegradable is a very interesting field in relation to food packaging.[44,45]

Materials of biological origin can be classified into nanomaterials, polymers, and biologically obtained fibers, as well as their composites (Figure 6.1). Specifically and of interest, biologically obtained polymers are found in four types: (1) obtained from biomass, (2) synthetics such as polylactic acid, (3) produced by microorganisms, and (4) biodegradable synthesized by petrochemical monomers. In recent years, studies based on bio-based polymeric resources cause impact in the food industry because it has been investigated that they can be implemented in the growth of plastics that are biodegradable.[46] In the food industry, a great trend has been observed toward bio-based reinforcements such as essential oils obtained from plants and natural additives for edible biofilms that help to enhance the oxygen barrier and likewise antibacterial and antifungal properties.[44] Based on this, bio-based food packaging materials to be implemented must meet some requirements such as permeability, antibacterial, and mechanical properties, taking into account that the application of these materials usually involves their composites.[46]

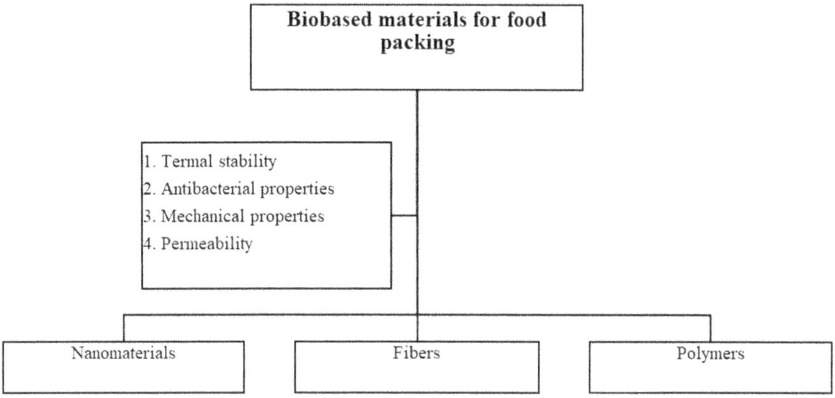


**FIGURE 6.1** General information about bio-based materials and their classification.

### *6.6.1 AGROINDUSTRIAL BY-PRODUCTS (BIOPOLYMERS)*

Obtaining biopolymers from agroindustrial by-products seems to be a very attractive initiative because it allows in an important way to reduce costs in the necessary materials that are implemented so that they can enter the market and thus compete with conventional plastics, especially in some applications that have a lower added cost (such as food packaging), and in this same area to be able to valorize macromolecules that are present in agroindustrial waste and thus give them important applications that enhance them in different fields of the food industry.[47] The incorporation of agroindustrial products in food processing and transformation has a very innovative trend in the study of biopolymers science with a clear and concrete outlook in the application of active food packaging. In recent years, some naturally obtained compounds have been raised to be able to be implemented in polymeric matrices with the main objective of helping in quality, in their application potential, and in a very important way food safety.[48] The trend of being able to implement some of the natural additives in polymeric formulations has boosted the growth toward a greater number of researches that are based on natural extracts obtained from plants, agricultural residues, or essential oils; the impact on their interest is thanks to their original compounds (flavonoids, proanthocyanidins, tannins, and phenolic acids) and thus to be applied in food packaging materials. Recent research has proposed some agroindustrial products for their proven composition such as *litchi chinensis*, specifically from the peel where flavonoids, polyphenols, among others, are found, and it is proposed that from them can be obtained antioxidant films of guar gum/caboxymethylcellulose and the implementation of its extract in oxygen-sensitive products.[49,50] Similarly, compounds found in *coffea arabica* such as polyphenols and flavonoids, as well as antioxidant activity, can be added to biopolymer matrices and thus have a great impact on their composition.[51,52] A final example in this section is the development from tomato peel that has a series of compounds that are interlinked with each other, esters or carboxylic acids that are known to be implemented in the development of biopolymers.[52] In case of humidity changes or dehydration, the peel of these three cases protects the pulp inside from microbial agents. Knowing these protective properties gives the idea to extract the compounds and implement them in the formation of biopolymers.

The main action of biopolymers is to help foods avoid oxidation, nutritional losses, damage by foodborne bacteria, presence of off-flavors, and loss of organoleptic quality due to the proliferation of microorganisms. The implementation of biopolymers obtained from gelatins, caseinates, and starches presents an interesting direction to develop sustainable materials. Biopolymers seem to be candidates for study to be transporters of bioactive compounds and to be applied in food packaging.[48]

### 6.6.2 EDIBLE BIOFILMS

Nowadays, there has been a great interest on the part of consumers in acquiring products that are very healthy, with high nutritional value, elaborated in a natural way and that have health benefits; all this is what has guided and driven research to develop biofilms and edible coatings so that they can be implemented in the food industry fulfilling these requirements.[53,54] The use of biofilms in the food industry seems to be a very accessible alternative, due to the issue of its food safety and the implementation of renewable resources, and therefore its use has an important impact on the environment, so they have been gradually improved to have the barrier protection in a proper way in different food products. With these important contributions in the industry, the industrialists mostly start to reduce in an impactful way the applications of plastics in food packaging, and therefore their focus is based on the development of innovative edible biofilms.[55,56]

Edible biofilms are considered edible products that form a thin layer on a solid surface with the strong characteristic of being a protective barrier with a semi-permeable characteristic toward gases and water vapor. Due to their high impact, these biofilms are used for the following main purposes: to extend the shelf life of foods, improve their mechanical properties, preserve the structure of the product they envelop, have the ability to act as a vehicle for bioactive compounds, and offer the possibility of improving food safety by transferring moisture, oxygen, and the compounds attributed to them are responsible for providing flavor, aroma, and color.[54] The main components of edible biofilms are polysaccharides, proteins, lipids, and resins, which are blended together to counteract some of the deficiencies that each of the individual components may have with the main objective of improving the properties of the materials and thus obtaining a better application.[57,58] Edible biofilms when applied to vegetables and fruits allow controlling respiration in a similar way to modified atmospheres, exerting a barrier to water vapor and gases, thus helping to reduce the deterioration of the fruit, therefore, being biofilms obtained naturally and having great benefits in the application to food, their use has been greatly enhanced.[57]

With the use of biofilms in food and their development as a new packaging material for coatings, many significant advantages over synthetic coatings have been reported, one of the most important being that they are edible and, in general, that they help to care for and respect the environment, being able to achieve a reduction of polymeric materials from nonrenewable sources, such as petroleum. It is important to consider the properties and characteristics of biofilms, since the type of packaging to be implemented should be defined and standardized according to the application required by the food.[53,54,59]

### 6.6.3 BIOADHESIVES

Bioadhesives can be defined as polymers coming from biological origin with adhesion and cohesion properties but under certain conditions of temperature, humidity, and surfaces.[9] According to the European Biomass Industry Association, a "biobased" product is defined as one that is composed entirely or partially of materials of biological origin.[49] Therefore, in recent years, the implementation and development of bioadhesives has increased significantly and materials such as starch, lignin, plant proteins, tannins, and plant oils have been investigated.[12,59] The application of these raw materials is with the main objective of acquiring substances that have the ability to provide the same characteristics of traditional adhesives such as adhesion performance (strength and types of chemical bonds between the surface and the adhesive), thermal stability, and water resistance.[9]

In the bioadhesives industry in recent years has been investigated the application of materials from renewable sources that are intended to have a special impact on the packaging sector and has had a significant growth thanks to the discovery of new biological molecules with adhesive properties that can be used in biological systems for the properties of biocompatibility and biodegradability, as well as the implementation of materials considered as waste from different industries to develop the potential of bioadhesives.[9,12]

### 6.6.4 NATURAL COLORIMETRIC INDICATORS OR SENSORS

The accelerated pace of people's lives is driving the need for fast, efficient, and direct testing of food quality and physiological processes to ensure proper protection against microbes, toxins, and bacteria. On the other hand, the increase of environmental contamination worldwide, the quality that assures the consumer that his food is safe, the control of food safety, as well as sanitary conditions, has acquired an important interest in researchers.[10] The means to assess food quality and therefore its immediate acquisition of the microbial load in the specific microenvironment of food can prevent spoiled food from being consumed, as well as provide information on some skin infections.[61] Therefore, it is of utmost importance to monitor microbial growth by simple and directly applicable methods for immediate identification of food condition without the need for pretreatment or complicated processes.

Microorganisms produce metabolic products that can alter the pH of this medium in which they inhabit and therefore multiply[61]; this has led to the development of high-performance pH indicators by visual colorimetric changes that may be suitable to show the presence of microbes containing qualitative or semiquantitative information that is attributed to the well-known colorimetric indicators in the addition of some substance that can turn to another color with changes in pH in polymers.[62,63] For the quantitative analysis of colorimetric sensors, basic visual technologies such as digital cameras with visualization software, analysis and processing of these images are used to differentiate between color, hue, or intensity measurements.[62] Despite all the high-tech methods, there is still a great demand for the development of new rapid, sensitive, and inexpensive detection methods, so there has been a greater focus on natural colorimetric indicators that are rationally used as rapid, sensitive, selective, and inexpensive colorimetric biosensors for the detection of some bacteria such as *Staphylococcus aureus*. These natural colorimetric indicators have at least one biological component that can be the substrate or the sensing element; therefore, they are applied mainly in food packaging, based on a colorimetric determination of the analyte.[3] The most striking research focuses on anthocyanidins, which in recent years with their implementation have been shown to act as pigment molecules in a natural way to color plants, vegetables, and fruits; it has been reported that they are stable in aqueous studies and thus give various colors that are based on the protonation and deprotonation of the structure in varied pH values.[3,63] For example, a pH sensor made from starch and potato agar with anthocyanidin extracts showed color variations at pH 2.0–10.0.[64] Table 6.1 shows some of the examples of natural colorimetric indicators developed to monitor food quality in industry in recent years.

Based on the composition of some agroindustrial products such as litchi chinensis and coffea arabica have aroused interest in the same way in possible candidates as pigments that are used as natural colorimetric indicators, which is suggested as future research to prove it.

## 6.7 BIO-BASED NANOMATERIALS

Nanotechnology is an interdisciplinary domain, encompassing a vast array of technological fields. It serves as an instrumental tool for the creation, examination, and manipulation of materials at the nanoscale, roughly

**TABLE 6.1** Natural Colorimetric Indicators Investigated for Food Control.

| | **Function** | **Food** | **Sensing material (source)** | **Responsense range/limit** | **References** |
|---|---|---|---|---|---|
| **Starch food** | Colorimetric sensing, antibacterial | Starch-PVA | Limonene and anthocyanin | pH 1–14 | [65] |
| | Colorimetric sensing | Corn starch—glycerol | Anthocyanin (blueberry powder) | pH 1–14 | [66] |
| | Colorimetric sensing antioxidant | Cassava starch—glycerol | Chlorophyll and carotenoids (green tea and basil) | pH 3 and pH 12 | [67] |
| **Chitosan matrix** | Colorimetric sensing | Chitosan | Anthocyanin (grapes) | pH 2–13 | [68] |
| | Colorimetric sensing antioxidant | Chitosan | Anthocyanin (blueberry and blackberry pomace) | pH 2,4,5,6,7, 10,12 | [69] |
| **Cellulose matrix** | Colorimetric sensing | Celullose (filter paper) | Anthocyanin (rose and red cabbage) | N/A | [70] |
| | Colorimetric sensing | Bacterial cellulose nanofibers | Anthocyanin (red cabbage) | pH 2–10 | [71] |

between 1 and 100 nm.[72] The emergence of innovative nanomaterials has opened avenues to improve food quality and safety, optimize agricultural growth, and enhance environmental monitoring. These materials are characterized by a large surface-to-volume ratio, and exhibit enhanced characteristics such as solubility, durability, diffusion capacity, and thermodynamic stability.[72,73]

One of the most promising avenues of nanotechnology in the agro-industrial realm is the development of bio-based packaging using bionanomaterials. Unlike conventional nondegradable plastic polymers, bio-based packaging incorporates qualities like biodegradability, bio-compatibility, and eco-friendliness, making it an ideal sustainable choice. Desirably, the biopolymers used for this purpose should be renewable, cost-effective, and preferably derived from waste materials.[72,73]

Various polysaccharides, including cellulose, chitosan, starch, pectin, alginate, carrageenan, pullulan, and kefiran, are widely researched for biodegradable coatings in food packaging.[72,73] These substances can form films with strong barrier properties against $O_2$ and $CO_2$ transmission. Their tensile strength is comparable to synthetic polymers, with high amylose starch films showing tensile strength similar to low-density polyethylene films. To further improve the barrier and structural properties of these films, the introduction of plasticizers has been investigated, which alter the films' mechanical traits, enhancing flexibility and resistance to breaking.[72]

Despite the benefits of bio-based materials, they often exhibit inferior barrier and structural properties compared to traditional fossil-based plastics. Their natural brittleness and low thermal resistance have so far limited their applications. This has led to significant research into nanotechnology to develop biomaterials with the necessary properties for diverse food product applications. Notably, biopolymer nanocomposites, multi-phase materials with a biopolymer matrix and nanoscale fillers, show promise in this area, potentially addressing the traditional shortcomings of bio-based materials.[74,75]

Current research in active biopolymer nanocomposites aims to enhance features such as oxygen barrier effectiveness. Introducing antimicrobial or antioxidant agents, along with nanofillers, is an effective approach to potentially match or even exceed the performance of standard plastic packaging in terms of food lifespan.[75]

Bio-based polymers, crucial in sustainable packaging, can be categorized based on their source and production process. For example, chitin, mainly found in insect and crustacean exoskeletons, is promising due to its abundance. Starch, derived from sources like wheat and rice, varies in its amylopectin and amylose content depending on the source, affecting its mechanical properties. Similarly, collagen, a protein rich in amino acids, has been used in films, and its derivative, gelatin, is known for excellent film-forming capabilities.[81]

In line with sustainable development, the pursuit of environmentally friendly, biodegradable, and sustainable food packaging materials has become a priority. Cellulose, Earth's most abundant biopolymer, is a promising candidate in this endeavor. Its potential in nanocellulose-based composites, with their high strength, large surface area, and superior barrier properties, highlights its significance in future food packaging solutions.[76]

Advancements in nanotechnology have paved the way for research and development of bionanomaterials for sustainable applications, especially within the food packaging realm. Based on biopolymers, these materials offer sought-after features such as biodegradability, biocompatibility, and eco-friendliness.[72]

Polysaccharides: These biopolymers are being intensively researched for their potential to form biodegradable coatings in food packaging. Among them are:

Cellulose: Earth's most abundant biopolymer. The promise of nanocellulose-based composites, with their high strength, extensive surface area, and excellent barrier properties, is significant for developing future food packaging solutions.[76–78]

Chitosan: A derivative of chitin, predominantly found in insect exoskeletons and crustaceans.

Starch: Sourced from wheat and rice, it varies in amylopectin and amylose composition, influencing its mechanical properties.[77,78, 81]

Other notable polysaccharides include pectin, alginate, carrageenan, pullulan, and kefiran. These substances can produce films with strong barrier properties against the movement of O2 and CO2.

Proteins: Apart from collagen and its derivative gelatin, other proteins like soy proteins, wheat gluten, and zein (corn protein) have been explored. These proteins can be transformed into biodegradable films for packaging due to their inherent structural properties.[76]

Lipids: These have been employed to enhance the barrier properties against water vapor. Natural waxes such as beeswax and plant-based waxes can be integrated into biopolymer films to reduce their permeability.[81]

To further improve the barrier and structural qualities of these films, the use of plasticizers has been explored, which modify the films' mechanical characteristics, enhancing flexibility and break resistance.[72]

The integration of nanotechnology with these biological materials has overcome the inherent limitations of bio-based polymers, such as brittleness and low thermal resistance, creating more diverse applications in the food packaging industry and promoting a sustainable future.

The convergence of nanotechnology with agricultural–industrial by-products offers transformative potential in food packaging.[79,80] Utilizing the unique properties of nanoscale materials combined with bio-based polymers, there is an opportunity to create sustainable, eco-friendly packaging solutions that meet both environmental goals and the functional requirements of the food industry.[81,82]

## 6.8 FUTURE PROSPECTIVES IN FOOD PACKING

In recent decades, the discernible environmental repercussions of plastics have become more pronounced. This awareness among consumers has catalyzed a heightened demand for sustainable packaging alternatives that can uphold shelf life standards without compromising product quality. Consequently, the food packaging sector has emerged as a vanguard in pioneering inventive solutions to address these pressing challenges.

Traditionally, edible packaging has served dual purposes, enhancing food aesthetics and preservation.[83] The growing interest in edible films has been spurred by its potential to partially substitute nonbiodegradable synthetic packaging materials. Notably, edible films derived from food-processing by-products have been garnering significant attention due to their economical and sustainable nature.[74,83] Such by-products are sourced from a plethora of industries, including those associated with fruits, vegetables, marine foods, and edible oils. Recent advances have delved into tailoring the properties of these films using composite formulations and nanotechnology applications, with cross-linking of various biopolymers showing promise.[83]

However, like many burgeoning technologies, edible films encounter obstacles. While attempts have been made to enhance the quality of these films, challenges persist, including inferior mechanical properties, insufficient resistance against external elements, and inadequate physical attributes.[83]

A paradigm shift is evident in the realm of food packaging, focusing on the nexus between material science innovations and food preservation. Edible packaging, driven by the development of various polymers, has emerged as a vanguard in environmentally friendly food packaging.[84,85] Emphasizing sustainability, there has been a discernible shift toward biopolymers and nanotechnology, the latter being instrumental in designing composite materials tailored for food packaging needs.[80,84]

Nanotechnology is poised to make significant inroads, considering safety concerns related to food packaging. The dynamism in R&D in this sphere is underscored by the emphasis on active and intelligent packaging, providing environmentally friendly alternatives.[86] Nanocellulose, for instance, derived from abundant banana biomass, has showcased its potential as a stellar biocomposite in food packaging.[87] It enhances the attributes of bioplastics, making them more amenable to practical applications.

While nanotechnology's potential in food packaging is clear, health concerns related to nanoscale particles, such as inhalation risks and bioaccumulation in the food chain, require thorough risk assessments.[80,88] Additionally, public awareness of health, safety, and environmental implications associated with nanotechnology in the food industry is critical.[88]

Future packaging decisions will rely on a comprehensive approach, considering the nature of the food item, its shelf life, and surrounding environmental conditions. As synthetic packaging faces environmental challenges, bio-packaging, which may eventually replace synthetic options, is on the rise. Polysaccharide-based materials, despite some drawbacks, are expected to transform the edible film and sensor segments in food packaging.[78]

The onset of the fourth industrial revolution brings with it the promise of virtual platforms like sensor systems, crucial for food and packaging product authentication and traceability.[89]

The dynamics of food packaging are undergoing a transformative shift, propelled by a confluence of sustainability imperatives and technological advancements. Although challenges persist, the collaborative endeavors of researchers, policymakers, and industry stakeholders instill confidence in a forthcoming era where food packaging seamlessly integrates sustainability with enhanced functionality. This synergy reflects a collective commitment to navigating the evolving landscape of packaging practices toward a more sustainable and resilient future.

## KEYWORDS


- **food packing**
- **edible films**
- **active food packing**
- **agroindustrial by-products**


## REFERENCES


1. Robertson, G. L. History of Food Packaging. In: *Reference Module in Food Science*; Elsevier, 2019. DOI: 10.1016/b978-0-08-100596-5.22535-3.
2. Nemes, S. A.; Szabo, K.; Cristian Vodnar, D. Coatings Applicability of Agro-Industrial By-Products in Intelligent Food Packaging. DOI: 10.3390/coatings10060550.
3. Halonen, N. et al. Bio-Based Smart Materials for Food Packaging and Sensors—A Review. *Front. Mater.* **2020,** *7*, 521914. DOI: 10.3389/FMATS.2020.00082/BIBTEX.
4. Dainelli, D.; Gontard, N.; Spyropoulos, D.; Zondervan-van den Beuken, E.; Tobback, P. Active and Intelligent Food Packaging: Legal Aspects and Safety Concerns. *Trends Food Sci. Technol.* **2008,** *19* (1), S103–S112. DOI: 10.1016/J.TIFS.2008.09.011.
5. FAO. *Appropriate Food Packaging Solutions for Developing Countries*; Food and Agriculture Organization of the United Nations: Rome, 2014.
6. Wikström, F.; Williams, H.; Trischler, J.; Rowe, Z. The Importance of Packaging Functions for Food Waste of Different Products in Households. *Sustainability* **2019,** *11* (9), 2641. DOI: 10.3390/SU11092641.
7. Nemes, S. A.; Szabo, K.; Vodnar, D. C. Applicability of Agro-Industrial By-Products in Intelligent Food Packaging. *Coatings* **2020,** *10* (6), 550. DOI: 10.3390/COATINGS 10060550.
8. Food Packaging and Shelf Life: A Practical Guide—Google Libros. https://books.google.com.mx/books?hl=es&lr=&id=t-LDcaqYD6EC&oi=fnd&pg=PP1&ots=UkjyYiY4YO&sig=SkmhumH32cNlbCKZ7Kjj1NMzYWU&redir_esc=y#v=onepage&q&f=false (accessed on 13 Nov 2023).
9. Zamudio, G.; Gosset, G.; Sabido-Ramos, A. Bioadhesivos: La transición hacia los compuestos bio-basados. **2021,** *25* (1).
10. Colorimetric Indicators Based on Anthocyanin Polymer Composites: A Review | Enhanced Reader.
11. Dehghani, S.; Hosseini, S. V.; Regenstein, J. M. Edible Films and Coatings in Seafood Preservation: A Review. *Food Chem.* **2018,** *240*, 505–513. DOI: 10.1016/J.FOODCHEM.2017.07.034.
12. Ferdosian, F.; Pan, Z.; Gao, G.; Zhao, B. Bio-Based Adhesives and Evaluation for Wood Composites Application. *Polymers* **2017,** *9* (2). DOI: 10.3390/polym9020070.

13. Ada, E.; Kazancoglu, Y.; Gozacan-Chase, N.; Altin, O. Challenges for Circular Food Packaging: Circular Resources Utilization. *Appl. Food Res.* **2023,** *3*, 100310. DOI: 10.1016/j.afres.2023.100310.
14. Alamri, M. S. et al. Food Packaging's Materials: A Food Safety Perspective. *Saudi J. Biol. Sci.* **2021,** *28*. DOI: 10.1016/j.sjbs.2021.04.047.
15. Rudra, S. G.; Kumar, P.; Mukherjee, A.; Ahari, H.; Soufiani, S. P. Smart and Active Food Packaging: Insights in Novel Food Packaging. 2021. DOI: 10.3389/fmicb.2021.657233.
16. Estevinho, M. et al. Molecules The Emergence of Edible and Food-Application Coatings for Food Packaging: A Review. 2022. DOI: 10.3390/molecules27175604.
17. Yuvaraj, D. et al. Advances in Bio Food Packaging—An Overview. 2021. DOI: 10.1016/j.heliyon.2021.e07998.
18. Shaikh, S.; Yaqoob, M.; Aggarwal, P. An Overview of Biodegradable Packaging in Food Industry. *Curr. Res. Food Sci.* **2021,** *4*, pp. 503–520. DOI: 10.1016/j.crfs.2021.07.005.
19. Vasile, C.; Baican, M. Molecules Review Progresses in Food Packaging, Food Quality, and Safety-Controlled-Release Antioxidant and/or Antimicrobial Packaging. 2021. DOI: 10.3390/molecules26051263.
20. Dash, K. K.; Deka, P.; Punia Bangar, S.; Chaudhary, V.; Trif, M.; Rusu, A. Applications of Inorganic Nanoparticles in Food Packaging: A Comprehensive Review. 2022. DOI: 10.3390/polym14030521.
21. Jildeh, N. B.; Matouq, M. Nanotechnology in Packing Materials for Food and Drug Stuff Opportunities. *J. Environ. Chem. Eng.* **2020,** *8*, 104338. https://doi.org/10.1016/J.JECE.2020.104338.
22. Yan, M. R.; Hsieh, S.; Ricacho, N. Innovative Food Packaging, Food Quality and Safety, and Consumer Perspectives. *Processes* **2022,** *10*, 747. https://doi.org/10.3390/pr10040747.
23. Leyva-Jiménez, F. J.; Oliver-Simancas, R.; Castangia, I.; Rodríguez-García, A. M.; Alañón, M. E. Comprehensive Review of Natural Based Hydrogels as an Upcoming Trend for Food Packing. *Food Hydrocoll.* **2023,** *135*, 108124. https://doi.org/10.1016/j.foodhyd.2022.108124.
24. Vasile, C.; Baican, M. Progresses in Food Packaging, Food Quality, and Safety—Controlled-Release Antioxidant and/or Antimicrobial Packaging. *Molecules* **2021,** *26*, 1263. https://doi.org/10.3390/molecules26051263.
25. Mousavi Khaneghah, A.; Hashemi, S. M. B.; Limbo, S. Antimicrobial Agents and Packaging Systems in Antimicrobial Active Food Packaging: An Overview of Approaches and Interactions. *Food Bioprod. Process.* **2018,** *111*, 1–19. https://doi.org/10.1016/J.FBP.2018.05.001.
26. Kuai, L.; Liu, F.; Chiou, B. Sen; Avena-Bustillos, R. J.; McHugh, T. H.; Zhong, F. Controlled Release of Antioxidants from Active Food Packaging: A Review. *Food Hydrocoll.* **2021,** *120*, 106992. https://doi.org/10.1016/J.FOODHYD.2021.106992.
27. Radoor, S.; Shivanna, J. M.; Baswaraj, S.; Chowdhury, C. R.; Jayakumar, A.; Karayil, J.; Kim, J. T.; Lee, J.; Siengchin, S. Lightweight and Sustainable Materials for Food Packaging Applications. In: *Lightweight and Sustainable Composite Materials*; Elsevier, 2023; pp 219–240. https://doi.org/10.1016/B978-0-323-95189-0.00010-X.
28. Wei, H.; Seidi, F.; Zhang, T.; Jin, Y.; Xiao, H. Ethylene Scavengers for the Preservation of Fruits and Vegetables: A Review. *Food Chem.* **2021,** *337*, 127750. https://doi.org/10.1016/J.FOODCHEM.2020.127750.

29. Duncan, T. V. Applications of Nanotechnology in Food Packaging and Food Safety: Barrier Materials, Antimicrobials and Sensors. *J. Colloid Interface Sci.* **2011,** *363*, 1–24. https://doi.org/10.1016/J.JCIS.2011.07.017.
30. Mustafa, F.; Reescu, S. Chemical and Biological Sensors for Food-Quality Monitoring and Smart Packaging. *Foods* **2018,** *7* 168. https://doi.org/10.3390/FOODS7100168.
31. Forghani, S.; Almasi, H.; Moradi, M. Electrospun Nanofibers as Food Freshness and Time-Temperature Indicators: A New Approach in Food Intelligent Packaging. *Innov. Food Sci. Emerg. Technol.* **2021,** *73*, 102804. https://doi.org/10.1016/J.IFSET.2021.102804.
32. Sohail, M.; Sun, D. W.; Zhu, Z. Recent Developments in Intelligent Packaging for Enhancing Food Quality and Safety. *Crit. Rev. Food Sci. Nutr.* **2018,** *58*, 2650–2662. https://doi.org/10.1080/10408398.2018.1449731.
33. Zheng, L.; Liu, L.; Yu, J.; Shao, P. Novel Trends and Applications of Natural PH-Responsive Indicator Film in Food Packaging for Improved Quality Monitoring. *Food Control* **2022,** *134*, 108769. https://doi.org/10.1016/J.FOODCONT.2021.108769.
34. Singh, B. P.; Shukla, V.; Lalawmpuii, H.; Kumar, S.; Bhanu, C.; Singh, P. Indicator Sensors for Monitoring Meat Quality: A Review. *J. Pharmacogn. Phytochem.* **2018,** *7*, 809–812.
35. Fathi, P.; Karmakar, N. C.; Bhattacharya, M.; Bhattacharya, S. Potential Chipless RFID Sensors for Food Packaging Applications: A Review. *IEEE Sens. J.* **2020,** *20*, 9618–9636. https://doi.org/10.1109/JSEN.2020.2991751.
36. Adley, C. C. Past, Present and Future of Sensors in Food Production. *Foods* **2014,** *3*, 491–510. https://doi.org/10.3390/FOODS3030491.
37. Ma, Q.; Du, L.; Wang, L. Tara Gum/Polyvinyl Alcohol-Based Colorimetric NH3 Indicator Films Incorporating Curcumin for Intelligent Packaging. *Sens. Actuators B Chem.* **2017,** *244*, 759–766. https://doi.org/10.1016/J.SNB.2017.01.035.
38. Maftoonazad, N.; Ramaswamy, H. Design and Testing of an Electrospun Nanofiber Mat as a PH Biosensor and Monitor the PH Associated Quality in Fresh Date Fruit (Rutab). *Polym. Test.* **2019,** *75*, 76–84. https://doi.org/10.1016/j.polymertesting.2019.01.011.
39. Haupt, K.; Mosbach, K. Molecularly Imprinted Polymers and Their Use in Biomimetic Sensors. *Chem. Rev.* **2000,** *100*, 2495–2504. https://doi.org/10.1021/CR990099W/ASSET/CR990099W.FP.PNG_V03.
40. Song, T.; Qian, S.; Lan, T.; Wu, Y.; Liu, J.; Zhang, H. Recent Advances in Bio-Based Smart Active Packaging Materials. *Foods* **2022,** *11*, 2228. https://doi.org/10.3390/FOODS11152228.
41. Tracey, C. T.; Predeina, A. L.; Krivoshapkina, E. F.; Kumacheva, E. A 3D Printing Approach to Intelligent Food Packaging. *Trends Food Sci. Technol.* **2022,** *127*, 87–98. https://doi.org/10.1016/J.TIFS.2022.05.003.
42. Zhou, W.; Fang, J.; Tang, S.; Wu, Z.; Wang, X. 3D-Printed Nanocellulose-Based Cushioning–Antibacterial Dual-Function Food Packaging Aerogel. *Molecules* **2021,** *26*, 3543. https://doi.org/10.3390/MOLECULES26123543.
43. Zeidler, H.; Klemm, D.; Böttger-Hiller, F.; Fritsch, S.; Le Guen, M. J.; Singamneni, S. 3D Printing of Biodegradable Parts Using Renewable Biobased Materials. *Procedia Manuf.* **2018,** *21*, 117–124. https://doi.org/10.1016/J.PROMFG.2018.02.101.

44. Asgher, M.; Qamar, S. A.; Bilal, M.; Iqbal, H. M. N. Bio-Based Active Food Packaging Materials: Sustainable Alternative to Conventional Petrochemical-Based Packaging Materials. *Food Res. Int.* **2020,** *137*. DOI: 10.1016/j.foodres.2020.109625.
45. Sehgal, R.; Sharma, V.; Gupta, R. Biobased Materials in Food Packaging. In: *Advanced Applications of Biobased Materials: Food, Biomedical, and Environmental Applications*; 2023; pp 147–172. DOI: 10.1016/B978-0-323-91677-6.00022-2.
46. Wang, J.; Euring, M.; Ostendorf, K.; Zhang, K. Biobased Materials for Food Packaging. *J. Bioresour. Bioprod.* **2022,** *7* (1), 1–13. DOI: 10.1016/j.jobab.2021.11.004.
47. Bojorges, H.; Mendez, D. A.; Pérez Bassart, Z.; Fabra Rovira, M. J.; Martínez Abad, A.; Lopez Rubio, A. Polímeros naturales obtenidos de residuos agroindustriales y sus potenciales aplicaciones. *Revista de plásticos modernos: Ciencia y tecnología de polímeros, ISSN 0034-8708, Vol. 122, N. 772, 2021 (Ejemplar dedicado a: Polímeros naturales, biodegradables y de fuentes renovables)* **2023,** *122* (772), 2. https://dialnet.unirioja.es/servlet/articulo?codigo=8169852&info=resumen&idioma=SPA (accessed on 11 Nov 2023).
48. Garrigós, M. C.; Mellinas, A. C.; Ramos, M.; Valdés, A. Sustainable Bioplastics María Carmen Garrigós et al Potential Natural Additives and Agro-Industrial By-Products in Biopolymer Formulations for Active Food Packaging. *J. Bioremediat. Biodegrad.* **2016,** *7* (6). DOI: 10.4172/2155-6199.C1.005.
49. Soni, R.; Agrawal, S. *Litchi chinensis*: Taxonomy, Botany and Its Cultivars. In: *Lychee Disease Management*; Springer: Singapore, 2017; pp 191–215. DOI: 10.1007/978-981-10-4247-8_12.
50. Castillo-Olvera, G.; Carrillo-Inungaray, M. L.; Reyes-Munguia, A.; Muñiz-Marquez, D. B.; Valencia-Hernández, L. J.; Wong-Paz, J. E. Litchi (*Litchi chinensis*): Generalidades, actividad biológica y aplicaciones. *TIP Revista Especializada en Ciencias Químico-Biológicas* **2022,** *25*. DOI: 10.22201/fesz.23958723e.2022.508.
51. Cunza, S. S.; Pillihuaman, E. A.; Roca, E. G. R.; Cunza, S. S.; Pillihuaman, E. A.; Roca, E. G. R. Actividad antioxidante, polifenoles y flavonoides de Coffea arabica de cinco regiones peruanas. *Revista de la Sociedad Química del Perú* **2020,** *86* (4), 343–354. DOI: 10.37761/RSQP.V86I4.307.
52. Wong-Paz, J. E.; Guyot, S.; Aguilar-Zárate, P.; Muñiz-Márquez, D.B.; Contreras-Esquivel, J. C.; Aguilar, C. N. Structural Characterization of Native and Oxidized Procyanidins (Condensed Tannins) from Coffee Pulp (*Coffea arabica*) Using Phloroglucinolysis and Thioglycolysis-HPLC-ESI-MS. *Food Chem.* **2021,** *340*. DOI: 10.1016/j.foodchem.2020.127830.
53. Dehghani, S.; Hosseini, S. V.; Regenstein, J. M. Edible Films and Coatings in Seafood Preservation: A Review. *Food Chem.* **2018,** *240*, 505–513. DOI: 10.1016/J.FOODCHEM.2017.07.034.
54. Fernández Valdés, D.; Bautista Baños, S.; Fernández Valdés, D.; Ocampo Ramírez, A.; García Pereira, A.; Falcón Rodríguez, A. Películas y recubrimientos comestibles: una alternativa favorable en la conservación poscosecha de frutas y hortalizas. *Revista Ciencias Técnicas Agropecuarias* **2015,** *24* (3), 52–57. http://scielo.sld.cu/scielo.php?script=sci_arttext&pid=S2071-00542015000300008&lng=es&nrm=iso&tlng=es (accessed on 11 Nov 2023).

55. Puscaselu, R.; Gutt, G.; Amariei, S. Rethinking the Future of Food Packaging: Biobased Edible Films for Powdered Food and Drinks. *Molecules* **2019,** *24* (17). DOI: 10.3390/MOLECULES24173136.
56. Enciso, N. A. A.; Enciso, N. A. A.; Hortua, N. S. Evaluación de películas comestibles de almidón de yuca y proteína aislada de soya en la conservación de fresas. *Nova* **2010,** *8* (14), 171–182. DOI: 10.22490/24629448.448.
57. Han, J. H. *Innovations in Food Packaging*; Elsevier Academic, 2005.
58. Ribeiro, C.; Vicente, A. A.; Teixeira, J. A.; Miranda, C. Optimization of Edible Coating Composition to Retard Strawberry Fruit Senescence. *Postharvest Biol. Technol.* **2007,** *44* (1), 63–70. DOI: 10.1016/j.postharvbio.2006.11.015.
59. Navarros, T. M. Efecto de la composicion de recubrimientos comestibles a base de hidroxiplimetilcelulosa y cera de abeja en la calidad de ciruelas, mandarinas y milojas. *Universidad Politécnica de Valencia. Departamento de tecnología de alimentos*, 2007.
60. Urushida, Y. et al. Identification and functional characterization of a novel barnacle cement protein. *FEBS J.* **2007,** *274* (16), 4336–4346. DOI: 10.1111/j.1742-4658.2007.05965.x.
61. Taormina, P. J. Purposes and Principles of Shelf Life Determination. In: *Food Safety and Quality-Based Shelf Life of Perishable Foods*; 2021; pp 1–26. DOI: 10.1007/978-3-030-54375-4_1.
62. Sani, M. A.; Azizi-Lalabadi, M.; Tavassoli, M.; Mohammadi, K.; McClements, D. J. Recent Advances in the Development of Smart and Active Biodegradable Packaging Materials. *Nanomaterials* **2021,** *11* (5), 1331. DOI: 10.3390/NANO11051331.
63. Genovese, M. E.; Zia, J.; Fragouli, D. Natural and Biocompatible Optical Indicators for Food Spoilage Detection. In: *Sustainable Food Packaging Technology*; 2021; pp 369–393. DOI: 10.1002/9783527820078.CH14.
64. Choi, I.; Lee, J. Y.; Lacroix, M.; Han, J. Intelligent pH Indicator Film Composed of Agar/Potato Starch and Anthocyanin Extracts from Purple Sweet Potato. *Food Chem.* **2017,** *218*, 122–128. DOI: 10.1016/J.FOODCHEM.2016.09.050.
65. Liu, B.; Xu, H.; Zhao, H.; Liu, W.; Zhao, L.; Li, Y. Preparation and Characterization of Intelligent Starch/PVA Films for Simultaneous Colorimetric Indication and Antimicrobial Activity for Food Packaging Applications. *Carbohydr. Polym.* **2017,** 157, 842–849. DOI: 10.1016/J.CARBPOL.2016.10.067.
66. Luchese, C. L.; Pavoni, J. M. F.; Spada, J. C.; Tessaro, I. C. Influence of Blueberry and Jaboticaba Agroindustrial Residue Particle Size on Color Change of Corn Starch Based Films Submitted to Different pH Values Solutions. *J. Renew. Mater.* **2019,** *7* (3), 235–243. DOI: 10.32604/JRM.2019.00033.
67. Medina-Jaramillo, C.; Ochoa-Yepes, O.; Bernal, C.; Famá, L. Active and Smart Biodegradable Packaging Based on Starch and Natural Extracts. *Carbohydr. Polym.* **2017,** *176*, 187–194. DOI: 10.1016/J.CARBPOL.2017.08.079.
68. Yoshida, C. M. P.; Maciel, V. B. V.; Mendonça, M. E. D.; Franco, T. T. Chitosan Biobased and Intelligent Films: Monitoring pH Variations. *LWT—Food Sci. Technol.* **2014,** *55* (1), 83–89. DOI: 10.1016/J.LWT.2013.09.015.
69. Kurek, M.; Hlupić, L.; Ščetar, M.; Bosiljkov, T.; Galić, K. Comparison of Two pH Responsive Color Changing Bio-Based Films Containing Wasted Fruit Pomace as a Source of Colorants. *J. Food Sci.* **2019,** *84* (9), 2490–2498. DOI: 10.1111/1750-3841.14716.

70. Ma, Y.; Li, L.; Wang, Y. Development of PLA-PHB-Based Biodegradable Active Packaging and Its Application to Salmon. *Pack. Technol. Sci.* **2018,** *31* (11), 739–746. DOI: 10.1002/PTS.2408.
71. Pourjavaher, S.; Almasi, H.; Meshkini, S.; Pirsa, S.; Parandi, E. Development of a Colorimetric pH Indicator Based on Bacterial Cellulose Nanofibers and Red Cabbage (*Brassica oleraceae*) Extract. *Carbohydr. Polym.* **2017,** *156*, 193–201. DOI: 10.1016/J.CARBPOL.2016.09.027.
72. Primožič, M.; Knez, Ž.; Leitgeb, M. Nanomaterials (Bio)Nanotechnology in Food Science-Food Packaging. 2021. DOI: 10.3390/nano11020292.
73. Fahmy, N. M. et al. Advances in Nanotechnology and Antibacterial Properties of Biodegradable Food Packaging Materials. 2020. DOI: 10.1039/d0ra02922j.
74. Sarfraz, J.; Gulin-Sarfraz, T.; Nilsen-Nygaard, J.; Pettersen, M. K. Nanomaterials Nanocomposites for Food Packaging Applications: An Overview. 2020. DOI: 10.3390/nano11010010.
75. Fotie, G.; Limbo, S.; Piergiovanni, L. Manufacturing of Food Packaging Based On Nanocellulose: Current Advances and Challenges. *Nanomaterials* **2020,** *10* (1726). DOI: 10.3390/nano10091726.
76. P. Emerging Food et al. Citation: Emerging Food Packaging Applications of Cellulose Nanocomposites: A Review. 2022. DOI: 10.3390/polym14194025.
77. Pal, K.; Sarkar, P.; Anis, A.; Wiszumirska, K.; Jarz̧ Ebski, M. Materials Polysaccharide-Based Nanocomposites for Food Packaging Applications. 2021. DOI: 10.3390/ma14195549.
78. Neši´c, A. N. et al. Molecules Prospect of Polysaccharide-Based Materials as Advanced Food Packaging. 2019. DOI: 10.3390/molecules25010135.
79. Tan, C. et al. Novel Bio-Based Materials and Applications in Antimicrobial Food Packaging: Recent Advances and Future Trends. *Int. J. Mol. Sci.* **2021,** *22*, 9663. DOI: 10.3390/ijms.
80. Kalpana Sastry, R.; Anshul, S.; Rao, N. H. Nanotechnology in Food Processing Sector—An Assessment of Emerging Trends. *J. Food Sci. Technol.* **2013,** *50* (5), 831–841. DOI: 10.1007/S13197-012-0873-Y.
81. Silva, F. A. G. S.; Dourado, F.; Gama, M.; Poças, F. Nanocellulose Bio-Based Composites for Food Packaging. *Nanomaterials* **2020**. DOI: 10.3390/nano10102041.
82. Cruz, R. M. S.; Silva, E. G. S.; Cardoso, S.; Bettencourt, A. F.; Ribeiro, I. A. C. Latest Trends in Sustainable Polymeric Food Packaging Films. 2022. DOI: 10.3390/foods12010168.
83. V, A. K.; Hasan, M.; Mangaraj, S.; M, P.; Verma, D. K.; Srivastav, P. P. Trends in Edible Packaging Films and Its Prospective Future in Food: A Review. *Appl. Food Res.* **2022,** *2* (1). DOI: 10.1016/j.afres.2022.100118.
84. Verma, M. K.; Shakya, S.; Kumar, P.; Madhavi, J.; Murugaiyan, J.; Rao, M. V. R. Trends in Packaging Material for Food Products: Historical Background, Current Scenario, and Future Prospects. *J. Food Sci. Technol.* **2021,** *58* (11), DOI: 10.1007/s13197-021-04964-2.
85. Biji, K. B.; Ravishankar, C. N.; Mohan, C. O.; Srinivasa Gopal, T. K. Smart Packaging Systems for Food Applications: A Review. *J. Food Sci. Technol.* **2015,** *52* (10), 6125–6135. DOI: 10.1007/S13197-015-1766-7/METRICS.

86. Majid, I.; Ahmad Nayik, G.; Mohammad Dar, S.; Nanda, V. Novel Food Packaging Technologies: Innovations and Future Prospective. *J. Saudi Soc. Agric. Sci.* **2018,** *17* (4), 454–462. DOI: 10.1016/j.jssas.2016.11.003.
87. Zaini, H. M et al. Banana Biomass Waste: A Prospective Nanocellulose Source and Its Potential Application in Food Industry—A Review. *Heliyon* **2023,** *9* (8). DOI: 10.1016/j.heliyon.2023.e18734.
88. Bajpai, V. K, et al. Prospects of Using Nanotechnology for Food Preservation, Safety, and Security. *ScienceDirect* **2018**. DOI: 10.1016/j.jfda.2018.06.011.
89. Chisenga, S. M.; Tolesa, G. N.; Workneh, T. S. Review Article Biodegradable Food Packaging Materials and Prospects of the Fourth Industrial Revolution for Tomato Fruit and Product Handling. 2020. DOI: 10.1155/2020/8879101.

## CHAPTER 7

# Green Nanoparticle Materials for a Sustainable Environment

JOBI J. MALAMEL and DHILNA FRANCIS

*Department of Zoology, Sacred Heart College (Autonomous), Cochin, Kerala, India*

## ABSTRACT

The rapid advancement of nanotechnology has brought forth numerous breakthroughs, yet concerns over the environmental impact of conventional nanoparticle synthesis methods have become increasingly evident. In response, this chapter explores the paradigm shift toward green nanoparticles—novel nanomaterials synthesized through ecofriendly approaches with minimal ecological footprints. The overarching goal is to provide an in-depth analysis of various green synthesis methods, the sustainability of resulting nanomaterials, and their diverse applications across industries. Challenges and future directions encapsulate the remaining hurdles in green nanoparticle synthesis, opportunities for further research, and the role of collaboration in advancing sustainable nanotechnology. The chapter emphasizes the transformative potential of ecofriendly approaches in shaping the future of nanoparticle synthesis and applications. It is aimed for a collective commitment to sustainable practices, urging researchers, industries, and policymakers to collaborate in realizing the full potential of green nanoparticles for a more environmentally conscious and responsible nanotechnological future.

Polymers and Functional Materials for a Cleaner Environment and Human Health: Sustainable Technologies and New Research. Ruby Varghese, Ann Rose Abraham, Shrikaant Kulkarni, & A. K. Haghi (Eds.)

## 7.1 INTRODUCTION

Nanotechnology has emerged as a groundbreaking field with vast potential for advancements in various industries. While these tiny structures offer promising applications in medicine, electronics, and materials science, their environmental impact has become a subject of increasing concern. Nanoparticles can enter the environment through natural processes, such as volcanic activity and forest fires, but human activities significantly contribute to their release. Industrial processes, manufacturing, and the use of nanomaterials in consumer products are primary sources of anthropogenic nanoparticles. Airborne nanoparticles are released during combustion processes, such as vehicle emissions and industrial activities. Combustion generates ultrafine particles, including nanoparticles, which can travel over long distances and pose respiratory risks to both humans and wildlife. In water, nanoparticles can be discharged through industrial effluents or enter the aquatic system through runoff from nanomaterial-containing products. Soil contamination can occur when nanoparticles are released into the environment and accumulate in the soil through deposition or wastewater irrigation. The potential environmental impact of nanoparticles encompasses a wide range of ecological and human health risks. In aquatic ecosystems, nanoparticles can adversely affect aquatic organisms, including fish, algae, and invertebrates. Studies have shown that certain nanoparticles can disrupt physiological processes, such as respiration and reproduction, in aquatic organisms. Additionally, the accumulation of nanoparticles in sediments may lead to long-term effects on benthic communities. The interaction between nanoparticles and soil microorganisms may alter microbial activity, affecting soil fertility and plant growth. The impact on plants can extend to the food chain, with potential consequences for herbivores and carnivores. The potential toxicity of certain nanoparticles to flora and fauna can disrupt ecosystems and threaten biodiversity. Aquatic organisms, in particular, may be vulnerable to the effects of nanoparticles, leading to population declines and imbalances within ecosystems. The use of nanomaterials as biosensors has currently a huge impact and a quick development within the usage of smart nanobiomaterials. Biosensors are accurate and offer a cost-effective approach for the detection of pathogenic infectious agents in the natural environment and also clinical specimens.[50]

As nanotechnology continues to advance, the integration of ecofriendly approaches becomes imperative for ensuring a sustainable and responsible future.[25] The potential benefits of nanoparticles must be realized without compromising the health of ecosystems and human well-being. Researchers, industries, and policymakers must collaborate to develop and implement environmentally conscious practices that mitigate the environmental impact of nanoparticles.[51] Through a collective commitment to ecofriendly approaches, we can harness the transformative power of nanotechnology while safeguarding the delicate balance of our planet's ecosystems. In an era marked by rapid technological advancements, the intersection of nanotechnology and environmental sustainability has given rise to the promising field of green nanotechnology. Traditional applications of nanotechnology have undeniably revolutionized various industries, offering unprecedented precision and efficiency. However, the environmental repercussions of conventional nanomaterials have raised concerns about their long-term impact on ecosystems and human health. Despite its transformative potential, the environmental footprint of conventional nanomaterials, characterized by resource-intensive synthesis processes and potential ecological hazards, has underscored the need for a more sustainable paradigm. Green Nanotechnology strives to address this by integrating principles of environmental responsibility and ecological harmony into the core of nanomaterial design and utilization. Adopting environmentally benign synthesis methods, minimizing waste production, and prioritizing biodegradability, Green Nanotechnology seeks to mitigate the negative ecological impact associated with traditional nanomaterials.[46] This paradigm shift is not only crucial for preserving biodiversity and ecosystem integrity but also aligns with global efforts to transition toward a circular and sustainable economy. This chapter is an attempt to contribute valuable insights to the growing discourse on sustainable nanotechnology and to pave the way for a more responsible and environmentally conscious approach to the utilization of nanomaterials, fostering a harmonious coexistence between technological progress and ecological well-being.

## 7.2 MICROBIAL SYNTHESIS

Nanoparticles exhibit unique physical, chemical, and biological properties distinct from their bulk counterparts (Figure 7.1). The synthesis of nanoparticles has traditionally involved chemical methods, often employing hazardous materials and energy-intensive processes. In recent years, there has been a growing interest in harnessing the capabilities of microorganisms, particularly bacteria and fungi, for the ecofriendly and cost-effective synthesis of nanoparticles.[44] Nanoparticle synthesis using bacteria and fungi has gained significant attention in recent years due to its ecofriendly and cost-effective nature. Both bacteria and fungi have the ability to reduce metal ions to form nanoparticles through a process known as bioreduction.

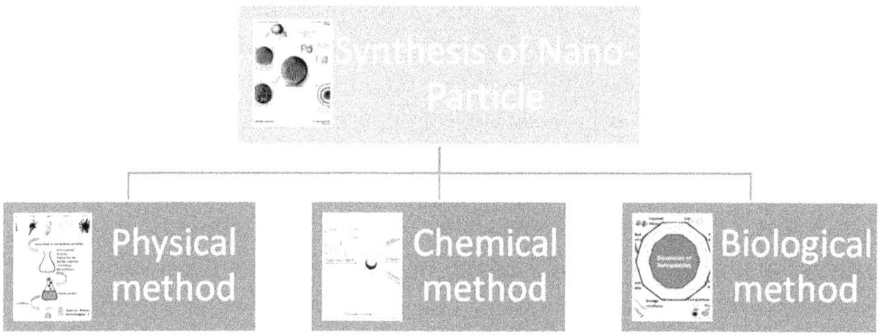


**FIGURE 7.1** Methods of nanoparticle synthesis.

### *7.2.1 PROCESS OF BACTERIAL AND FUNGAL NANOPARTICLE SYNTHESIS*

Selection of Bacterial Strain: Choose a bacterial strain known for its ability to reduce metal ions. Common strains include *Bacillus*, *Escherichia coli* (*E. coli*), and Pseudomonas.

Bacterial Culture Preparation: Cultivate the selected bacterial strain under optimal conditions, including temperature, pH, and nutrient medium, to promote growth.

Metal Ion Precursor Addition: Introduce a solution containing metal ions, usually in the form of metal salts (e.g., silver nitrate for silver nanoparticles), into the bacterial culture.

Bioreduction: Bacteria have enzymes or proteins that facilitate the reduction of metal ions. This reduction process occurs either on the bacterial cell surface or within the periplasmic space.

Nanoparticle Nucleation: Reduced metal ions nucleate to form nanoparticles. The characteristics of the nanoparticles depend on the bacterial strain and the synthesis conditions.

Stabilization: Biomolecules produced by bacteria, such as proteins or polysaccharides, act as stabilizing agents for the synthesized nanoparticles, preventing agglomeration.

Harvesting and Purification: Separate the bacterial biomass and culture medium from the nanoparticle solution. Further purification steps may be employed to obtain a more concentrated and pure nanoparticle solution.

Characterization: Characterize the synthesized nanoparticles using techniques like UV-Vis spectroscopy to confirm formation, transmission electron microscopy (TEM) to visualize size and shape, and X-ray diffraction (XRD) to determine crystalline structure.[47,48]

#### *7.2.1.1 FUNGAL NANOPARTICLE SYNTHESIS*

Selection of Fungal Strain: Choose a fungal strain known for its ability to secrete enzymes, such as reductases, involved in the reduction of metal ions. Common strains include Aspergillus, Fusarium, and Candida.

Fungal Culture Preparation: Cultivate the selected fungal strain under appropriate conditions, including temperature, pH, and nutrient medium (Figure 7.2).

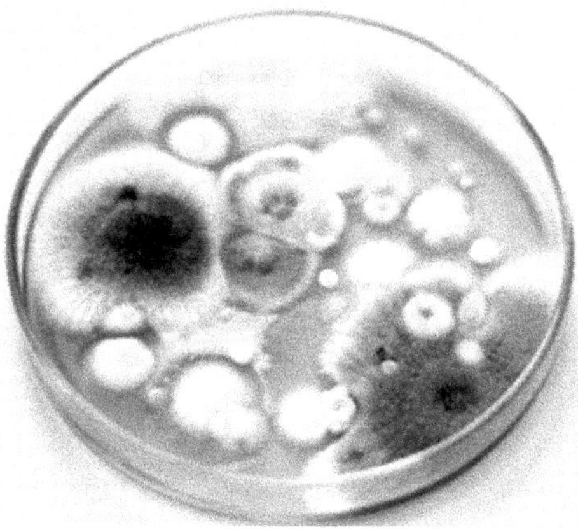

**FIGURE 7.2** Fungal culture.

Metal Ion Precursor Addition: Introduce a solution containing metal ions into the fungal culture. Fungal synthesis often occurs extracellularly.

Bioreduction: Fungi secrete enzymes, such as reductases, which facilitate the reduction of metal ions to form nanoparticles outside the fungal cells.

Nanoparticle Formation: The reduced metal ions undergo nucleation to form nanoparticles. The characteristics of these nanoparticles depend on the fungal strain and synthesis conditions (Figure 7.3).

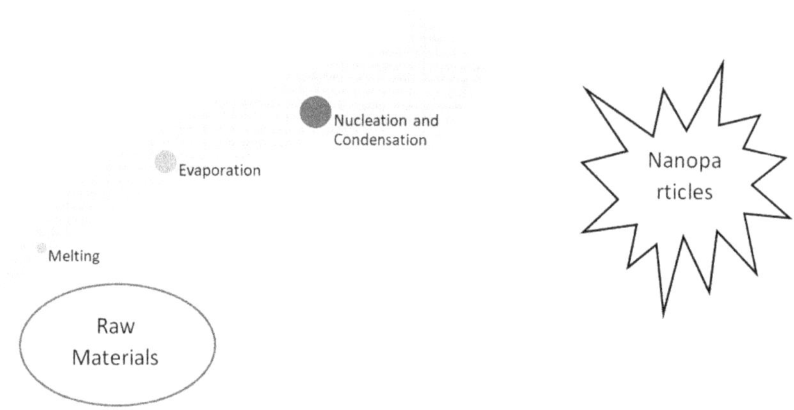


**FIGURE 7.3** Nanoparticle formation.

Stabilization: Similar to bacterial synthesis, biomolecules produced by fungi play a role in stabilizing the nanoparticles and preventing agglomeration.

Harvesting and Purification: Separate the fungal biomass and culture medium from the nanoparticle solution. Additional purification steps may be implemented for concentrated and pure nanoparticles.

Characterization: Characterize the synthesized nanoparticles using various techniques, including UV-Vis spectroscopy, TEM, and XRD.[23]

These are the steps involved in the synthesis of bacterial and fungal particles (Figure 7.4). The significance of microorganism-mediated nanoparticle formation lies in its potential for sustainable and controlled synthesis, offering applications in fields such as medicine, materials science, and environmental remediation.

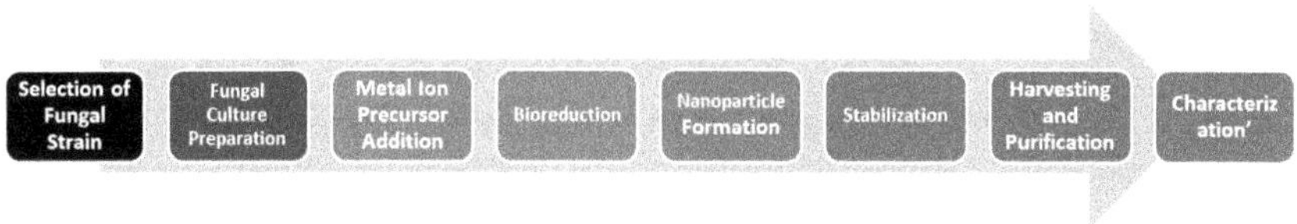


**FIGURE 7.4** Process of bacterial and fungal nanoparticle synthesis.

### 7.2.2 THE ROLE OF MICROORGANISMS IN GREEN NANOPARTICLE SYNTHESIS

Nanoparticles, with their unique properties and applications, have become pivotal in various fields such as medicine, electronics, catalysis, and environmental science. While traditional methods of nanoparticle synthesis often involve harsh chemicals and energy-intensive processes, there is a growing recognition of the potential for green synthesis methods, utilizing the remarkable capabilities of microorganisms. This essay explores the role of microorganisms, particularly bacteria and fungi, in green nanoparticle synthesis and the implications for sustainable and environmentally friendly nanotechnology.

#### 7.2.2.1 BACTERIAL GREEN NANOPARTICLE SYNTHESIS

Bacteria, being ubiquitous and versatile microorganisms, play a significant role in green nanoparticle synthesis. One of the key advantages of using bacteria is their ability to reduce metal ions to form nanoparticles through biological processes. Strains such as Bacillus, Escherichia coli, and Pseudomonas have been extensively studied for their efficacy in reducing various metal ions, including silver, gold, and platinum. The process typically involves the cultivation of bacteria under controlled conditions, followed by the introduction of metal ion precursors into the bacterial culture. The enzymes or proteins present on the bacterial cell surface or within the periplasmic space facilitate the reduction of metal ions, leading to the nucleation and formation of nanoparticles. The use of bacterial biomass also acts as a stabilizing agent, reducing the need for additional stabilizers and surfactants.

#### 7.2.2.2 FUNGAL GREEN NANOPARTICLE SYNTHESIS

Fungi, with their ability to secrete enzymes and thrive in diverse environments, provide another avenue for green nanoparticle synthesis. Species like Aspergillus, Fusarium, and Candida have demonstrated the capacity to reduce metal ions extracellularly, resulting in the formation of nanoparticles outside the fungal cells. Similar to bacterial synthesis, the process involves cultivating fungi under suitable conditions and introducing metal ion precursors into the fungal culture. The secreted enzymes, such as reductases, play a crucial role in the bioreduction of metal ions, leading to the creation of stable nanoparticles. The extracellular nature of fungal synthesis simplifies the separation of nanoparticles from the biomass, contributing to the ecofriendly aspect of the process.[38]

The role of microorganisms in green nanoparticle synthesis represents a significant advancement in the pursuit of sustainable nanotechnology. Bacteria and fungi, with their natural ability to reduce metal ions and create stable nanoparticles, offer a greener alternative to traditional synthesis methods. As we continue to unlock the potential of microbial processes, the convergence of microbiology and nanotechnology holds the promise of revolutionizing the way we produce nanoparticles, ensuring a more sustainable and environmentally conscious future.

### 7.2.3 ENVIRONMENTAL BENEFITS AND APPLICATIONS OF MICROBIAL GREEN NANOPARTICLE SYNTHESIS

Ecofriendly and Sustainable: Green nanoparticle synthesis using microorganisms eliminates the need for hazardous chemicals and energy-intensive processes, contributing to a more sustainable and environmentally friendly approach.

Cost-effective: Microbial synthesis is often more cost-effective compared to traditional chemical methods as it requires minimal additional reagents and energy inputs.

Biocompatibility: Nanoparticles synthesized through microbial processes are often coated with biomolecules, enhancing their biocompatibility and reducing potential toxicity concerns in medical applications.

Versatility: Microorganisms are highly adaptable, and the synthesis process can be optimized for a wide range of metal ions and nanoparticle types, providing versatility in nanoparticle production. The role of microorganisms in green nanoparticle synthesis represents a significant advancement in the pursuit of sustainable nanotechnology (Figure 7.5).

**FIGURE 7.5** Environmental benefits and applications of bacterial and fungal nanoparticle synthesis.

## 7.3 SUSTAINABLE NANOMATERIALS

The intersection of nanotechnology and sustainability has led to the emergence of sustainable nanomaterials, representing a paradigm shift toward environmentally friendly and socially responsible practices in the field of nanoscience. Sustainable nanomaterials are designed and produced with the aim of minimizing adverse environmental impacts, conserving resources, and ensuring long-term societal benefits. Sustainable nanomaterial research is an interdisciplinary field that combines principles of nanotechnology, materials science, chemistry, biology, and environmental science to develop innovative solutions that balance technological progress with environmental and social responsibility.

### *7.3.1 DEFINITION AND CHARACTERISTICS*

Definition: Sustainable nanomaterials can be broadly defined as nanoscale substances or structures that are intentionally engineered to possess environmentally benign attributes throughout their life cycle – from synthesis and use to disposal. These materials are developed with a commitment to ecological balance, resource efficiency, and a reduced ecological footprint. The pursuit of sustainability in nanomaterials encompasses ethical considerations, social impacts, and the responsible use of technology for the benefit of present and future generations. Sustainable nanomaterials maintain or enhance the functional properties that make nanomaterials desirable, such as unique mechanical, optical, or electrical properties. This ensures that they can be effective substitutes for traditional materials without compromising performance.

#### 7.3.1.1 CHARACTERISTICS OF SUSTAINABLE NANOMATERIALS

Environmentally Benign Synthesis: Sustainable nanomaterials are synthesized using methods that minimize the use of hazardous substances, reduce energy consumption, and generate fewer byproducts. Green synthesis methods, involving biological entities or benign chemicals, are often employed.

Resource Efficiency: The production of sustainable nanomaterials prioritizes resource efficiency by minimizing raw material consumption and waste generation. Recycling and upcycling approaches are incorporated to extend the lifespan of materials and reduce environmental impact.

Biodegradability and Low Persistence: Sustainable nanomaterials are designed to exhibit biodegradability and low persistence in the environment. This characteristic ensures that the materials do not accumulate and pose long-term risks to ecosystems.

Nontoxicity and Biocompatibility: The design of sustainable nanomaterials includes considerations for nontoxicity and biocompatibility. This characteristic is particularly important for applications in medicine, where nanomaterials may interact with biological systems.

Renewable Resources: Whenever possible, sustainable nanomaterials are derived from renewable resources. This reduces dependency on finite resources and contributes to a more sustainable and circular economy.

Social Responsibility: Sustainable nanomaterials adhere to ethical and social responsibility standards. This includes fair labor practices, consideration of human rights, and the avoidance of practices that may lead to social inequality.

Functional Performance: While sustainability is a primary consideration, sustainable nanomaterials do not compromise on functional performance. They are engineered to meet specific application requirements and provide comparable or enhanced performance compared to traditional materials.

Regulatory Compliance: Sustainable nanomaterials adhere to regulatory frameworks governing nanotechnology and environmental protection.

Compliance with established guidelines ensures that these materials are safe for use and do not pose risks to human health or the environment.

Open Communication and Transparency: Transparency in communication regarding the characteristics, risks, and benefits of sustainable nanomaterials is a key feature. This fosters informed decision-making and responsible use in both scientific and industrial applications.

Sustainable nanomaterials represent a conscientious approach to harnessing the benefits of nanotechnology while minimizing its potential negative impacts. By integrating environmental, social, and economic considerations into the design and production processes, sustainable nanomaterials pave the way for a more responsible and resilient nanotechnological future. The ongoing development and adoption of these materials contribute to the broader goal of achieving sustainable and equitable technological advancements.

### *7.3.2 APPLICATIONS IN VARIOUS INDUSTRIES*

Sustainable nanomaterials find applications across diverse industries, offering innovative solutions to address challenges while minimizing environmental impact (Figure 7.6).

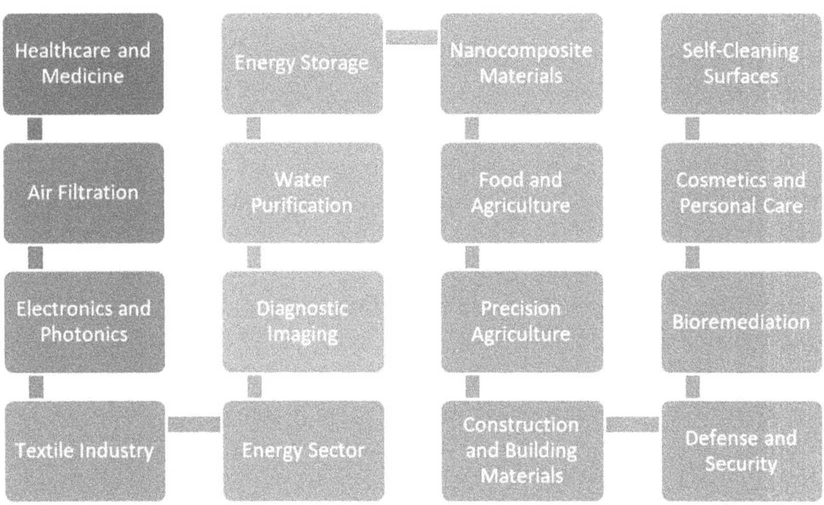


**FIGURE 7.6** Applications of bacterial and fungal nanoparticle synthesis in various industries.

Healthcare and Medicine (Drug Delivery Systems): Nanoparticles enable targeted drug delivery, improving treatment efficacy and reducing side effects.

Diagnostic Imaging: Nanomaterials enhance contrast agents for advanced imaging techniques, aiding in early disease detection.

Water Purification: Nanomaterials are used for removing pollutants and contaminants from water, contributing to sustainable and efficient water treatment.

Air Filtration: Nanoparticles assist in developing efficient air filtration systems, capturing pollutants, and improving air quality.

Energy Sector (Solar Cells): Nanomaterials enhance the efficiency of solar cells, improving the conversion of sunlight into electricity.

Energy Storage: Nanomaterials are utilized in batteries and supercapacitors, enhancing energy storage capacity and lifespan.

Electronics and Photonics: Nanomaterials enable the development of flexible and lightweight electronic devices, contributing to the advancement of wearable technologies.

Textile Industry: Nanomaterials are incorporated into textiles to provide functionalities such as stain resistance, UV protection, and antimicrobial properties.

Nanocomposite Materials: Nanoparticles enhance the mechanical and thermal properties of textile composites for applications in the aerospace and automotive industries.

Food and Agriculture (Food Packaging): Nanomaterials improve the barrier properties of packaging materials, extending the shelf life of food products.

Precision Agriculture: Nanosensors and nanofertilizers enhance nutrient delivery and monitoring in agriculture, promoting sustainable farming practices.[4]

Construction and Building Materials: Nanoparticles improve the strength and durability of concrete, leading to more resilient and sustainable construction materials.

Self-Cleaning Surfaces: Nanocoatings with self-cleaning properties are applied to building surfaces, reducing maintenance and cleaning requirements.

Cosmetics and Personal Care (Sunscreen Formulations): Nanoparticles provide effective UV protection without the use of traditional, potentially harmful UV filters.

Antiaging Products: Nanomaterials contribute to the development of skincare products with improved penetration and targeted delivery.

Bioremediation: Nanomaterials assist in the remediation of contaminated sites through enhanced microbial degradation. Nanoparticles serve as contrast agents in advanced imaging techniques for detailed visualization in biological systems.[16]

Defense and Security (Sensor Technologies): Nanomaterial-based sensors enable sensitive and selective detection of various chemicals, aiding in security and defense applications.

Lightweight and Strong Materials: Nanocomposites improve the strength and resilience of materials used in armor and protective gear.

Waste Management: Nanomaterials contribute to more efficient waste-to-energy processes by enhancing catalytic reactions. Recycling and waste reduction efforts benefit from the development of nanocomposite materials with improved properties[14] (Table 7.1).

**TABLE 7.1** Diverse Applications of Nanoparticles in Industries.

| Industry | Example |
|---|---|
| Water Purification | Silver nanoparticles for antimicrobial filtration |
| Bioremediation | Fungal-synthesized nanoparticles for pesticide degradation |
| Biofuels | Enzyme-coated nanoparticles for bioconversion efficiency |
| Solar Energy | Plasmonic gold nanoparticles for light trapping and energy conversion |
| Biomedical | Magnetic iron oxide nanoparticles for targeted drug delivery |
| Food and Agriculture | Silver and zinc nanoparticles for antifungal and antibacterial protection |
| Cosmetics | Titanium dioxide nanoparticles for sun protection and UV absorption |
| Textiles | Nanocoated fabrics for water and oil repellency |
| Construction | Carbon nanotubes for concrete reinforcement and tensile strength |
| Electronics | Graphene-based nanoparticles for transparent and flexible electronics |
| Environmental Sensing | Biocompatible sensors for real-time monitoring |
| Catalysis | Enzyme-immobilized nanoparticles for green chemistry |
| Energy Storage | Graphene and metal oxide nanoparticles for energy density and power |
| Aerospace | Nanocomposites for structural integrity and thermal resistance |

### 7.3.3 ENVIRONMENTAL IMPACT ASSESSMENT (EIA) OF SUSTAINABLE NANOMATERIALS

Environmental Impact Assessment (EIA) of sustainable nanomaterials is crucial to ensure that the potential environmental risks and benefits associated with their production, use, and disposal are thoroughly understood. The assessment process involves evaluating various stages of the nanomaterial life cycle and considering factors such as toxicity, exposure, and persistence.[2]

Toxicity and Ecotoxicity Assessments: Examine the potential toxic effects of nanomaterials on human health and the environment. Assess the toxicity of nanomaterials in vitro and in vivo and evaluate their impact on various ecosystems and organisms.

Exposure Assessment: Understand how nanomaterials may be released into the environment and the pathways through which exposure can occur. Analyze the likelihood and extent of exposure during production, use, and end-of-life disposal. Consider factors such as release mechanisms and potential transformations.

Fate and Transport Studies: Determine how nanomaterials move through environmental compartments such as air, water, and soil.

Considerations: Assess the persistence, mobility, and transformation of nanomaterials in different environmental matrices to understand their long-term behavior.

Risk Assessment: Integrate information from toxicity, exposure, and fate studies to estimate the potential risks associated with nanomaterials. Evaluate risk levels and identify potential mitigation measures. Consider uncertainties and variability in data.

Regulatory Compliance: Ensure that the production and use of nanomaterials comply with existing regulatory frameworks. Assess whether nanomaterials meet regulatory requirements for safety, labeling, and reporting. Identify gaps in regulation and propose necessary adjustments.

Societal Impact Assessment: Consider the social implications of nanomaterial use, including human perceptions, ethical considerations, and social equity. Evaluate the potential societal benefits and concerns related to the adoption of sustainable nanomaterials. Consider stakeholder engagement and public perception.

Monitoring and Surveillance Programs: Establish systems for ongoing monitoring and surveillance to track the environmental behavior of nanomaterials. Develop methods for detecting and quantifying nanomaterials in environmental samples. Implement routine monitoring to assess any changes in environmental concentrations.

Life Cycle Cost Analysis (LCCA): Assess the economic aspects associated with the life cycle of nanomaterials. Evaluate costs related to production, use, maintenance, and end-of-life disposal. Compare the economic viability of sustainable nanomaterials with alternative materials or processes.

Communication and Transparency: Ensure transparent communication of assessment findings to stakeholders and the public. Communicate the results of the environmental impact assessment, including uncertainties and areas for improvement. Facilitate an open dialogue with stakeholders to address concerns and gather feedback.

Conducting a comprehensive environmental impact assessment for sustainable nanomaterials is essential for fostering responsible development and ensuring the safe and sustainable integration of nanotechnology into various industries. Regular updates and reviews of assessments should be conducted to incorporate new scientific knowledge and technologies.

## 7.4 CHALLENGES AND FUTURE DIRECTIONS IN GREEN NANOPARTICLE SYNTHESIS

While green nanoparticle synthesis offers numerous advantages, it is not without its challenges. Addressing these challenges is crucial for the widespread adoption of environmentally friendly nanoparticle synthesis methods.[24]

### *7.4.1 REMAINING CHALLENGES IN GREEN NANOPARTICLE SYNTHESIS*

Control over Size and Shape: Achieving precise control over the size and shape of nanoparticles remains a challenge in green synthesis methods. The variability in biological systems may lead to a broad range of particle sizes and shapes, which can impact the desired properties.

Reproducibility: Green synthesis methods often involve biological entities, and the inherent variability in these systems can lead to challenges in reproducibility. Consistency in nanoparticle size, shape, and properties is essential for many applications.

Scale-up Issues: Transitioning from laboratory-scale synthesis to largescale production can be challenging. Green synthesis methods may face scalability issues, and optimizing conditions for largescale production while maintaining environmentally friendly practices is a complex task.

Purity and Contamination: Green synthesis methods may introduce impurities or contaminants from biological sources. Purifying nanoparticles without compromising their green attributes can be challenging and requires careful optimization.

Biocompatibility and Toxicity: Despite being considered environmentally friendly, nanoparticles synthesized through green methods must still undergo rigorous testing for biocompatibility and potential toxicity. Understanding the biological interactions and long-term effects is crucial for applications in medicine and other fields.

Stability and Agglomeration: Green-synthesized nanoparticles may exhibit lower stability compared to those produced through traditional methods. Preventing agglomeration and ensuring the stability of nanoparticles in various environments are ongoing challenges.

Standardization of Methods: The lack of standardized protocols for green nanoparticle synthesis poses a challenge for researchers and industries. Standardization is essential for comparing results across studies and ensuring consistency in nanoparticle properties.

Economic Viability: Green synthesis methods may face economic challenges in terms of cost-effectiveness compared to traditional methods. The availability and cost of biological resources and the optimization of processes for efficiency are important considerations.

Regulatory Approval: Obtaining regulatory approval for green-synthesized nanoparticles can be challenging due to the unique characteristics of these particles. Developing standardized protocols for toxicity testing and ensuring compliance with regulatory requirements are ongoing efforts.

Public Perception: Public perception of green-synthesized nanoparticles and their safety may influence acceptance and adoption. Public awareness and understanding of the benefits and risks associated with green nanoparticle synthesis are crucial for its successful integration into various applications (Figure 7.7).

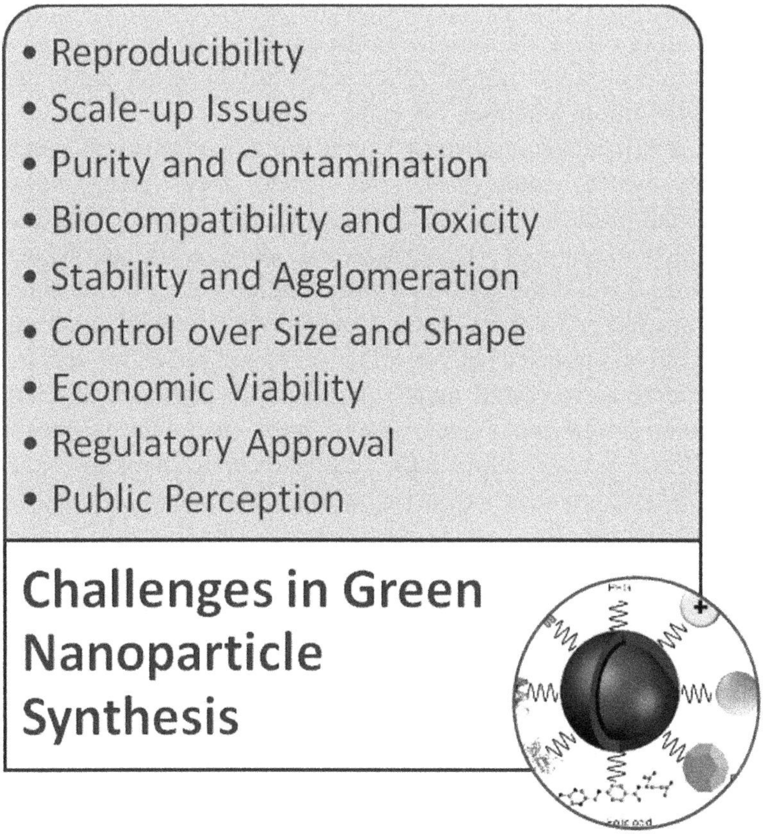


**FIGURE 7.7** Challenges in green nanoparticle synthesis.

### *7.4.2 OPPORTUNITIES FOR FURTHER RESEARCH AND INNOVATION FOR GREEN NANOTECHNOLOGY*

Green nanotechnology offers exciting opportunities for further research and innovation across various domains. In material synthesis, researchers can explore novel green chemistry approaches and sustainable feedstocks to create nanomaterials with reduced environmental impact. Emphasis can be placed on developing biodegradable nanomaterials for applications in packaging and medical fields, where environmentally friendly alternatives are urgently needed. Additionally, the integration of green nanomaterials in clean energy technologies, such as solar cells and energy storage devices, presents a promising avenue for addressing climate change. Researchers can focus on optimizing nanomaterial properties to enhance energy conversion and storage efficiency while ensuring the materials' sustainability throughout their life cycle.

Another critical research area lies in the development of nanotechnology for water treatment and purification. Investigating advanced nanomaterials with superior adsorption and catalytic properties can lead to more efficient removal of contaminants from water sources. Moreover, the creation of smart agricultural systems incorporating nanotechnology holds significant promise for sustainable farming practices. By designing nanomaterials for targeted delivery of nutrients and pesticides, researchers can contribute to increased agricultural productivity while minimizing the environmental impact associated with conventional farming methods. Overall, these research directions in green nanotechnology have the potential to revolutionize industries, mitigate environmental challenges, and contribute to a more sustainable and resilient future.

Life Cycle Analysis and Sustainability Metrics: Conduct comprehensive life cycle analyses for green-synthesized nanoparticles to assess their overall sustainability. Develop metrics for evaluating the environmental and societal impact of these nanoparticles throughout their life cycle.[33]

### *7.4.3 THE ROLE OF COLLABORATION IN ADVANCING GREEN NANOTECHNOLOGY*

Collaboration plays a pivotal role in advancing green nanotechnology, fostering interdisciplinary research, and addressing complex challenges associated with the development and application of environmentally friendly nanomaterials. The synergy between researchers, industry professionals, policymakers, and other stakeholders is essential for accelerating the progress of green nanotechnology.[49]

Multidisciplinary Research, Interdisciplinary Teams: Collaboration brings together experts from diverse fields, including chemistry, biology, materials science, engineering, and environmental science. This multidisciplinary approach is crucial for a comprehensive understanding of the complexities involved in green nanotechnology.

Knowledge Exchange, Academic–Industry Collaboration: Collaboration between academia and industry facilitates the exchange of knowledge, enabling researchers to gain insights into real-world applications, industry requirements, and market dynamics. Industry partners, in turn, benefit from cutting-edge research and innovative solutions.

Shared Resources and Infrastructure, Pooling Resources: Collaborative efforts allow for the sharing of resources, equipment, and infrastructure. This is particularly important in nanotechnology, where specialized equipment and facilities may be required for synthesis, characterization, and testing.

Addressing Complex Challenges, Holistic Problem-Solving: Green nanotechnology involves addressing complex challenges related to synthesis, characterization, toxicity, scalability, and sustainable applications. Collaboration enables a holistic approach to problem-solving, drawing on the expertise of researchers with complementary skills.[30]

Technology Transfer, Bridging the Gap: Collaboration facilitates the transfer of technology and knowledge from research institutions to industry. This bridge between academia and industry accelerates the translation of research findings into practical applications, products, and processes.

Policy Development, Engaging Policymakers: Collaborating with policymakers ensures that the development of green nanotechnology aligns with regulatory frameworks and environmental standards. Policymakers can benefit from the expertise of researchers in understanding the potential risks and benefits of nanomaterials.

International Collaboration, Global Knowledge Exchange: Green nanotechnology is a global endeavor, and international collaboration enhances the exchange of ideas, best practices, and research findings. Collaborative projects across borders contribute to a more comprehensive understanding of environmental impacts and regulatory considerations.[31]

Educational Initiatives, Training the Next Generation: Collaboration supports educational initiatives by providing opportunities for students to engage in interdisciplinary research. This helps train the next generation of scientists and professionals who can contribute to the sustainable development of nanotechnology.[52]

Standardization and Guidelines, Creating Consensus: Collaboration is essential for the development of standardized protocols, guidelines, and best practices in green nanotechnology. Consensus among researchers, industry partners, and regulatory bodies ensures a unified approach to safety and sustainability.

Stakeholder Engagement, Inclusive Decision-Making: Collaboration involves engaging with various stakeholders, including researchers, industry representatives, environmental groups, and the public. Inclusive decision-making processes ensure that diverse perspectives are considered in the development and deployment of green nanotechnologies.[43]

Collaboration acts as a catalyst for advancing green nanotechnology by promoting interdisciplinary research, facilitating technology transfer, addressing challenges comprehensively, and ensuring that innovations align with sustainability goals and societal needs. The collective expertise and shared resources within collaborative networks contribute to the responsible development and application of environmentally friendly nanomaterials.[42]

## 7.5 CONCLUSIONS

Nanotechnology has emerged as a transformative force with the potential to revolutionize various industries. As the field advances, the ecological imperative of green nanotechnology has gained prominence, urging researchers and practitioners to integrate environmentally sustainable practices. In the concluding chapter of "Looking Ahead: Building a Sustainable Tomorrow," the emphasis lies on understanding and addressing the ecological implications of nanotechnology, steering it toward a future that prioritizes environmental responsibility.

Green nanotechnology is a paradigm shift that recognizes the need for sustainable solutions in the development and application of nanomaterials. It seeks to mitigate the environmental impact of nanotechnology processes and products, acknowledging the interconnectedness of technological progress and ecological well-being.[29] The imperative is not merely a call for restraint but a mandate to innovate responsibly, ensuring that the benefits of nanotechnology do not come at the expense of the environment.

In the quest for a sustainable tomorrow, the conclusion of this discourse underscores the urgency of incorporating green principles into nanotechnology practices. This imperative arises from the recognition that conventional approaches may inadvertently contribute to ecological degradation. Nanomaterials, with their unique properties, have the potential to bring about positive changes, but their unintended consequences demand a re-evaluation of methodologies.

Green nanomaterials are engineered to fulfill specific functions while minimizing environmental harm. This proactive design approach considers the principles of green chemistry, focusing on the synthesis of nanomaterials with reduced toxicity, biodegradability, and low energy requirements. Such conscientious design choices pave the way for sustainable nanotechnology applications that resonate with the ecological imperative.

In the pursuit of sustainability, it is crucial to have a collaborative effort among researchers, policymakers, industry stakeholders, and the public. Establishing interdisciplinary platforms for dialogue and knowledge exchange fosters a shared understanding of the ecological imperatives in nanotechnology. It also facilitates the development of regulations and guidelines that ensure responsible innovation without stifling progress.

This chapter envisions a future where nanotechnology becomes synonymous with environmental stewardship. It enlists the support of educational institutions, encouraging the incorporation of green nanotechnology principles into curriculum by nurturing a generation of scientists and engineers with a deep-rooted commitment to ecological imperatives to perpetuate sustainable practices throughout the evolution of nanotechnology.

In short, the ecological imperative of green nanotechnology encapsulates a profound call to action. It urges a departure from traditional paradigms and a reimagining of nanotechnology as a catalyst for positive environmental change. The chapter resonates with the belief that by integrating green principles into the core of nanotechnology, we can not only unlock its full potential but also ensure a sustainable and harmonious coexistence with the planet. It beckons us to look ahead, envisioning a future where the marvels of nanotechnology unfold hand-in-hand with the preservation of our ecological heritage.

## KEYWORDS

- **bioremediation**
- **collaboration**
- **ecofriendly**
- **ecological footprint**
- **future**
- **paradigm shift**
- **toxicity**

## REFERENCES


1. Abdelbasir, S. M.; Shalan, A. E. An Overview of Nanomaterials for Industrial Wastewater Treatment. *Korean J. Chem. Eng.* **2019,** *36* (8), 1209–1225.
2. Abdel-Shafy, H. I.; Mansour, M S. M. Solid Waste Issue: Sources, Composition, Disposal, Recycling, and Valorization, Egypt. *J. Pet.* **2018,** *27*, 1275–1290.
3. Allison, R. P. Electrodialysis Treatment of Surface and Waste Waters. Technical Paper, GE Water and Process Technologies Paper TP1032EN, 2013. http://www.gewater.com/pdf/Technical%20Papers_Cust /Americas/English/ TP103EN.pdf
4. Afrouzi, Y. M.; Marzbani, P.; Omidvar, A. The Effect of Moisture Content on the Retention and Distribution of Nano-Titanium Dioxide in the Wood. *Maderas Ciencia y Tecnología* **2015,** *17*, 385–390.
5. Amil Usmani, M.; Khan, I. H.; Bhat, A. S.; Pillai, R.; Ahmad, N. K.; Mohamad Haafiz, M.; Oves, M. Current Trend in the Application of Nanoparticles for Waste Water Treatment and Purification: A Review. *Curr. Org. Synth.* **2017,** *14*, 206–226.
6. Arshad, A. Bacterial Synthesis and Applications of Nanoparticles. Nano. Sci. *Nano. Technol.* **2017,** *11* (2), 119.
7. Moshe, B.-S.; Lu, X.; Nejati, S.; Jaramillo, H.; Elimelech, M. In Situ Surface Functionalization of Reverse Osmosis Membranes with Biocidal Copper Nanoparticles. *Desalination* **2016,** *388* (2016), 1–8.
8. Baiocco, D.; Lavecchia, R.; Natali, S.; Zuorro, A. Production of Metal Nanoparticles by Agro-Industrial Wastes: A Green Opportunity for Nanotechnology. *Chem. Eng. Trans.* **2016,** *47* (6), 67–72.
9. Bhattacharyya, A.; Duraisamy, P.; Govindarajan, M.; Buhroo, A. A.; Prasad, R. Nano-Biofungicides: Emerging Trend in Insect Pest Control. *Adv. Appl. Through Fungal Nanobiotechnol.* **2016,** *1*, 307–319.
10. Cao, X.; Huang, X.; Liang, P.; Xiao, K.; Zhou, Y. A New Method for Water Desalination Using Microbial Desalination Cells. *Environ Sci. Technol.* **2009,** *43* (18), 7148–7152.
11. Cath, T. Y.; Childress, A. E.; Elimelech, M. Forward Osmosis: Principles, Applications, and Recent Developments. *J. Membr. Sci.* **2006,** *281*, 70–87.
12. Chattopadhyay, D.; Patel, B. Nano Metal Particles: Synthesis, Characterization and Application to Textiles. *Manuf. Nanostruct.* **2014,** *2*, 184–215.
13. Chhipa, H.; Joshi, P. Nanofertilisers, Nanopesticides and Nanosensors in Agriculture. In: *Nanoscience in Food and Agriculture*, Vol. 1; Springer: Cham, 2016; pp 247–282.
14. Chauhan, M.; Mittal, A.; Batra, J.; Philip, L. Nanoparticles for the Remediation of Environmental Pollutants: A Review. *J. Dispersion Sci. Technol.* **2017,** *38* (12), 1748–1761.
15. Ciambelli, P.; La Guardia, G.; Vitale, L. Nanotechnology for Green Materials and Processes. In: *Studies in Surface Science and Catalysis*, Vol. 179; Elsevier, 2019; pp 97–116.
16. Crawford, R. L.; Crawford, D. L. (2005). Bioremediation of Contaminated Soils: A Review of Microbial Processes. *Soil Sci. Soc. Am. J.* **2005,** *69* (2), 509–523.
17. Dahl, J. A.; Maddux, B. L. S.; Hutchison, J. E. Toward Greener Nanosynthesis. *Chem. Rev.* **2007,** *107*, 2228–2269.

18. Das, S.; Sen, B.; Debnath, N. Recent Trends in Nanomaterials Applications in Environmental Monitoring and Remediation. *Environ. Sci. Pollut. Res.* **2015,** *22* (23), 18333–18344.
19. Das, S.; Chakraborty, J.; Chatterjee, S.; Kumar, H. Prospects of Biosynthesized Nanomaterials for the Remediation of Organic and Inorganic Environmental Contaminants. *Environ. Sci. Nano* **2018,** *5* (12), 2784–2808.
20. Dimkpa, C. O. Can Nanotechnology Deliver the Promised Benefits Without Negatively Impacting Soil Microbial Life? *J. Basic Microbiol.* **2014,** *54*, 889–904.
21. Duan, H.; Wang, D.; Li, Y. Green Chemistry for Nanoparticle Synthesis. *Chem. Soc. Rev.* **2015,** *44* (16), 5778–5792.
22. Gomathi, T.; Rajeshwari, K.; Kanchana, V.; Sudha, P. N.; Parthasarathy, K. Impact of Nanoparticle Shape, Size, and Properties of the Sustainable Nanocomposites. In: *Sustainable Polymer Composites and Nanocomposites*; Inamuddin, T. S., Mishra, R. K.et al., Eds.; Springer, Cham, 2019.
23. Garcia, M. R.; Nguyen, H. T.; Kim, Y. J. Fungal-Mediated Green Synthesis of Nanoparticles: Mechanisms and Applications. *J. Fungal Res.* **2021,** *25* (3), 112–130.
24. Hemalatha, P.; Kumar, A. R.; Lekshmi, M. C.; Anilkumar, N. R. Challenges in Green Synthesis of Nanoparticles: A Review. *Renew. Sustain. Energy Rev.* **2018,** *89*, 742–760.
25. Holban, A. M.; Gestal, M. C.; Grumezescu, A. M. New Molecular Strategies for Reducing Implantable Medical Devices Associated Infections. *Curr. Med. Chem.* **2014,** *21*, 3375–3382.
26. Iwamoto, T.; Nasu, M. Current Bioremediation Practice and Perspective. *J. Biosci. Bioeng.* **2001,** *92*, 1.
27. Kah, M.; Hofmann, T. Nanopesticide Research: Current Trends and Future Priorities. *Environ. Int.* **2014,** *63*, 224–235.
28. Kale, A. P.; Gawade, S. N. Studies on Nanoparticle Induced Nutrient Use Efficiency of Fertilizer and Crop Productivity. *Green Chem. Technol. Lett.* **2016,** *2*, 88–92.
29. Kookana, R. S.; Boxall, A. B.; Reeves, P. T.; Ashauer, R.; Beulke, S.; Chaudhry, Q.; Cornelis, G.; Fernandes, T. F.; Gan, J.; Kah, M.; Lynch, I. Nanopesticides: Guiding Principles for Regulatory Evaluation of Environmental Risks. *J. Agric. Food Chem.* **2014,** *62*, 4227–4240.
30. Kulkarni, A. A.; Aminabhavi, T. M. Green Synthesis of Metal and Metal Oxide Nanoparticles: Challenges and Conclusions. *Green Chem.* **2018,** *20* (9), 1928–1943.
31. Le Berre, L.; Porter, A. L.; Rafols, I. International Collaboration in Nanotechnology Research: The Case of France and the United States. *Scientometrics* **2007,** *71* (1), 45–71.
32. Li, X.; Xu, H.; Chen, Z.; Guofang, C. Biosynthesis of Nanoparticles by Microorganisms and Their Applications. *J. Nanomater.* **2011,** Article ID 270974, 1–16.
33. Pokrajac, L.; Abbas, A.; Chrzanowski, W.; Dias, G. M.; Eggleton, B. J.; Maguire, S.; Maine, E.; Malloy, T.; Nathwani, J.; Nazar, L.; Sips, A.; Sone, J.; van den Berg, A.; Weiss, P. S.; Mitra, S. Nanotechnology for a Sustainable Future: Addressing Global Challenges with the International Network4Sustainable Nanotechnology. *ACS Nano* **2021,** *15* (12), 18608–18623.
34. Lombi, E.; Wenzel, W. W.; Adriano, D. C. Soil Contamination, Risk Reduction and Remediation. *Land Contamin. Reclam.* **1998,** *6* (4), 183–197.
35. Mukhopadhyay, S. S. Nanotechnology in Agriculture: Prospects and Constraints. *Nanotech. Sci. Appl.* **2014,** *7*, 63.

36. Pallavi Mehta, C. M.; Srivastava, R.; Arora, S.; Sharma, A. K. Impact Assessment of Silver Nanoparticles on Plant Growth and Soil Bacterial Diversity. *3 Biotech* **2016,** *6*, 254.
37. (a) Pan, Z. Food Decontamination Using Nanomaterials. *MOJ Food Process Technol.* **2015,** *1*, 00011. (b) Pandey, G. Challenges and Future Prospects of Agri-Nanotechnology for Sustainable Agriculture in India. *Environ. Technol. Innov.* **2018,** *11*, 299–307.
38. Patel, S.; Gupta, R.; Sharma, A. The Role of Microorganisms in Green Nanoparticle Synthesis: Mechanisms and Applications. *J. Microb. Nanotechnol.* **2022,** *10* (2), 87–104.
39. Patil, S. S.; Shedbalkar, U. U.; Truskewycz, A.; Chopade, B. A.; Ball, A. S. Nanoparticles for Environmental Clean-Up: A Review of Potential Risks and Emerging Solutions. *Environ. Technol. Innov.* **2016,** *5*, 10–21.
40. Prasad, R.; Kumar, V.; Prasad, K. S. Nanotechnology in Sustainable Agriculture: Present Concerns and Future Aspects. *Afr. J. Biotech.* **2014,** *13*, 705–713.
41. Prasad, R.; Bhattacharyya, A.; Nguyen, Q. D. Nanotechnology in Sustainable Agriculture: Recent Developments, Challenges, and Perspectives. *Front. Microbiol.* **2017,** *8*, 1–2.
42. Rajendran, S.; Arivalagan, P.; Rajasekar, A. Green Nanotechnology: Advancements and Challenges. *J. Clean. Prod.* **2019,** *236*, 117–135.
43. Schroeder, D.; Stone, V.; van den Hoven, J. Nanotechnology for Sustainability: What Does Nanotechnology Offer to Address Complex Sustainability Challenges? *Nanotechnol. Sci. Appl.* **2012,** *5*, 1–20.
44. Sharma, D.; Kanchi, S.; Bisetty, K. Biogenic Synthesis of Nanoparticles: A Review. *Arab. J. Chem.* **2019,** *12* (8), 3576–3600.
45. Singh, P.; Kim, Y. J.; Zhang, D.; Yang, D. C. Biological Synthesis of Nanoparticles from Plants and Microorganisms. *Trends Biotechnol.* **2016,** *34*, 588–599.
46. Smith, J. A.; Doe, J. B. Green Nanotechnology: A New Hope for Medical Biology. *J. Green Nanotechnol.* **2019,** *7* (2), 112–125.
47. Smith, J.; Johnson, A. B.; Chen, X. Sustainable Nanomaterials: Advances and Challenges. *ACS Sustain. Chem. Eng.* **2020,** *8* (15), 5603–5617.
48. Smith, A. B.; Johnson, C. D.; Patel, M. R. Exploring the Mechanisms: Bacterial Nanoparticle Synthesis from a Biological Perspective. *J. Nanopart. Res.* **2020,** *18* (5), 215–230.
49. Sujitha, I.; Kannan, S. Green Nanotechnology: A Review on Green Synthesis of Silver Nanoparticles—An Ecofriendly Approach. *Resour.—Efficient Technol.* **2018,** *4* (4), 508–520.
50. Veigas, B.; Fernandes, A. R.; Baptista, P. V. Aunps for Identification of Molecular Signatures of Resistance. *Front. Microbiol.* **2014,** *5*, 455.
51. Vo-Dinh, T. *Nanotechnology in Biology and Medicine: Methods, Devices, and Applications*; CRC Press, Taylor and Francis Group, 2007; p 762.
52. Warner, C. H.; Parkinson, D.; Collins, T. J. Green Nanotechnology Education: Towards Designing a Sustainable Future. *J. Clean. Prod.* **2017,** *162*, 1403–1412.

# CHAPTER 8

# Advancements in Catalytic Technologies for Promoting Sustainable Healthy Environment and Green Transport

AGAMPODI SANDALI INUPAMA DE ZOYSA, YASHWANTH NARAYANA, BHASKAR GAONKAR, and ROOPASHREE R.

*Department of Chemistry and Biochemistry, Jain Deemed to be University, Bengaluru, India*

## ABSTRACT

This chapter provides a comprehensive overview of ongoing research into utilizing nanomaterials for the catalytic conversion of harmful substances within vehicle exhaust control systems. These systems include diesel oxidation catalysts, diesel particulate filters, three-way catalysts, selective catalytic reduction units, and lean $NO_x$ traps. Nanoengineered materials introduce a wide array of opportunities for enhancing the catalytic performance of conventional catalysts, including precious metals like Pt, Pd, Rh, and Ru, as well as perovskites and base metal oxides such as V, Cu, Mn, Cr, Fe, and Mo. Innovative coating techniques are emerging as effective tools to elevate the catalytic oxidation of pollutants like CO, hydrocarbons (HCs), and the soluble organic fraction found in diesel particulate matter. These advancements facilitate the transformation of these pollutants into environmentally benign compounds like carbon dioxide and water. Furthermore, they contribute to the regeneration of particulate filters in diesel systems, aiding in the removal of soot buildup.

Polymers and Functional Materials for a Cleaner Environment and Human Health: Sustainable Technologies and New Research. Ruby Varghese, Ann Rose Abraham, Shrikaant Kulkarni, & A. K. Haghi (Eds.)

## 8.1 INTRODUCTION

Problems such as environmental, health, and ecological issues require strict order against the emission of toxic and harmful gases. These rules encompass restrictions on nitrogen oxide levels, particulate matter emissions, carbon monoxide output, hydrocarbon discharges, and the exhaust emissions from vehicle engines.[1] Greenhouse gases like $CO_2$ are facing growing scrutiny as well.[2] The exhaust emissions from diesel engines contain various powerful carcinogens associated with particulate matter, which also has a significant impact on the environment. Additionally, black carbon absorbs heat in the atmosphere, contributing to climate warming.[3] Correlations have been identified between increased levels of ambient fine particulate air pollution and elevated risks of mortality from both cardiopulmonary conditions and lung cancer.[4] Catalytic technologies are of significant importance in the realms of fuel processing, the chemical industry, and various environmental applications. Prominent instances of catalytic processes include the transformation of crude oil,[5] the manufacturing of diverse chemical goods,[6] and the implementation of regulations and technologies for controlling vehicular emissions.[7] Catalysts are vital elements within proton exchange fuel cells designed for automotive use, aimed at replacing traditional internal combustion engines.[8] To advance catalytic technologies, there is a requirement to create novel high-value products using cost-effective catalysts that reduce or substitute precious metals. Additionally, it involves the implementation of efficient and environmentally friendly conversion methods. One of the primary applications of catalytic technologies is the utilization of catalytic converters to remove harmful additives from vehicle exhaust emissions.[9] Following the era of the three-way catalyst (TWC), it served as the primary emission control mechanism for light-duty gasoline vehicles over the past three decades.[10] The diesel particulate filter (DPF) for the removal of particulate matter has emerged as the most crucial emission control device for vehicles.[11,12] Diesel Oxidation Catalysts (DOCs) are highly effective at catalytically converting carbon monoxide (CO), hydrocarbons (HCs), and the soluble organic portion of diesel particulate matter into carbon dioxide and water.[13] There are two fundamental distinctions between diesel engine exhaust and that of spark-ignited engines.[14] The primary differences between diesel engine exhaust and the exhaust from spark-ignited engines are twofold. Firstly, diesel exhaust is characterized by significantly higher levels of particulate matter and nitrogen oxide emissions. Secondly, diesel exhaust is notably leaner, meaning it contains far fewer unburned hydrocarbons (HCs) and carbon monoxide (CO) compared to the typical exhaust from gasoline engines. In such scenarios, two catalytic converters are employed to reduce nitrogen oxide emissions: the Selective Catalytic Reduction (SCR) system[15] and the Lean Nitrogen Oxide Trap (LNT) catalytic converter.[16,17]

The notable advancements made in recent years in the development and exploration of novel functional materials and nanoscale systems present promising prospects for their practical utilization across a wide range of human endeavors.[18–20] The alteration of properties in nanomaterials, such as their mechanical strength, electrical characteristics, and creative potential, is primarily driven by two key factors: the relative surface area and the influence of quantum effects.[21] From a catalytic technology perspective, nanotechnology opens up various avenues for enhancing catalyst efficiency. Numerous catalysts are composed of precious metals in particle form, and decreasing their size results in reduced metal usage. Furthermore, the inherent catalytic characteristics of the material undergo alterations due to a blend of localized coordination effects, shifts in surface size distribution, surface relaxation, and quantum size effects.[22–24] Recently, Cuenya and Behafrid conducted an extensive review that specifically delves into the impact of nanoparticle size and shape on chemisorption, as well as the catalytic properties, including factors like the initiation temperature of reactions, catalytic activity, and selectivity of nanocatalysts.[25]

Within this chapter, we conduct a comprehensive assessment of the ongoing advancements in the creation of nanostructured catalysts designed for vehicle exhaust control systems. This examination encompasses the scrutiny of existing experimental findings and the primary technical facets related to advanced catalytic technologies employed for the after-treatment of vehicle exhaust gases.

## 8.2 DIESEL OXIDATION CATALYST

The Diesel Oxidation Catalyst (DOC) plays a pivotal role in the diesel exhaust purification system, typically positioned immediately after the engine. Its primary role encompasses two key functions: (1) converting hydrocarbons (HCs) and carbon monoxide (CO) into water ($H_2O$) and carbon dioxide ($CO_2$). This serves the dual purpose of either reducing emissions originating from the engine or generating exothermic heat used for the regeneration of a Diesel Particulate Filter (DPF) and (2) oxidizing nitrogen oxides (NO) into nitrogen dioxide ($NO_2$), which is then utilized for the ongoing oxidation of soot particles on a Diesel Particulate Filter.[9,26] The apparatus comprises a cordierite (ceramic) honeycomb monolith system that does not rely on filters and possesses exceptional strength owing to its low ceramic porosity. However, this monolith's low porosity makes it unsuitable for distributing the catalyst. Consequently, a high-surface-area material known as a "wash coat" is affixed to the monolith's surface, and the catalytic metal is subsequently dispersed onto this wash coat, forming a layer approximately 20–40 μm thick.[13] The majority of oxidation catalysts found in the global market typically feature a ratio of platinum (Pt) to palladium (Pd) at approximately 5:2, with a standard loading range of 1.5 to 2.5 $kg/m^3$.[14,27] Pd is introduced to inhibit the coalescence of Pt particles, thereby enhancing the efficacy of Pt catalysts by fostering the creation of a Pt–Pd alloy.[28] Figure 8.1 represents the flow of the diesel oxidation catalyst.

**FIGURE 8.1** Schematic view of flow through the diesel oxidation catalyst.

The widely acknowledged reaction mechanism for the oxidation of CO and HC on Pt involves what is known as the Langmuir-Hinshelwood dual-site mechanism. In this mechanism, the critical step that determines the reaction rate is the surface interaction between adsorbed CO and HC molecules, which occur in close proximity to adsorbed oxygen.[29,30] You can find the illustration depicting the process of CO and HC oxidation on Pt nanoparticles in Figure 8.2.[31] Catalytic NO oxidation takes place on the surface of noble metal catalysts. Although NO is thermodynamically unstable and could potentially dissociate into oxygen and nitrogen, this dissociation does not occur because precious metal catalysts possess a strong affinity for oxygen. This affinity hinders the adsorption of NO, which is a crucial step in the NO reduction mechanism.[13,32] A DOC equipped with a substantial loading of catalytic material and a dense cell structure has the capability to physically capture and subsequently oxidize the volatile constituents of PM.[33] Present research endeavors are focused on enhancing the catalytic performance of the DOC. More specifically, researchers have synthesized highly dispersed core-shell and alloyed Pt–Pd bimetallic nanostructures supported on alumina and silica through a sequential and robust electrostatic adsorption process.[34] A range of diesel oxidation catalysts featuring varying Pt/V ratios were produced on a Ce–Zr–O support material through the impregnation technique.[35] The novel catalyst exhibited enhancements in both its catalytic efficiency and its resistance to sulfur. Notably, the most optimal catalytic performance was achieved with a 1 wt % V loading of the Pt nanoparticles. Additionally, the Ce–Zr–O nanostructured catalyst demonstrated active CO oxidation reactions taking place on its surface.[36] Catalysts consisting of Pt nanoparticles supported on a $ZrO_2$-modified $SiO_2$ surface exhibit outstanding capabilities in catalyzing the oxidation of CO. They showcase impressive low-temperature reactivity, hydrothermal durability, and resilience to sulfur exposure, making them highly promising for various applications.[37] The application of a $ZrO_2$ layer on $SiO_2$ surfaces before Pt impregnation results in the excellent dispersion and hydrothermal stability of Pt nanoparticles. This is attributed to the robust interaction between Pt and $ZrO_2$ phases. Additionally, a newly developed catalyst, featuring Au supported on $\gamma$-$Al_2O_3$ modified with both Fe and La, displays remarkable catalytic efficacy and enduring stability for CO oxidation reactions in both dry and moist air environments. The existence of Fe–La species enhances the stability of gold nanoparticles, ensuring their more secure deposition on the alumina surface.[38]

## 8.3 DIESEL PARTICULATE FILTER

Diesel exhaust particulate matter primarily comprises carbonaceous particles, commonly referred to as "soot". These soot particles originate within internal combustion engines due to the incomplete combustion of fuel. Soot can be divided into two categories: a) Soot exists in the form of fractal-like agglomerates made up of primary particles, which typically measure 15–30 nm in diameter. These agglomerates are primarily composed of carbon and contain traces of metallic ash. They are also coated with condensed, heavier-end organic compounds and sulfates. b) Nucleation particles consist of condensed hydrocarbons and sulfates, as illustrated in Figure 8.3.[39]

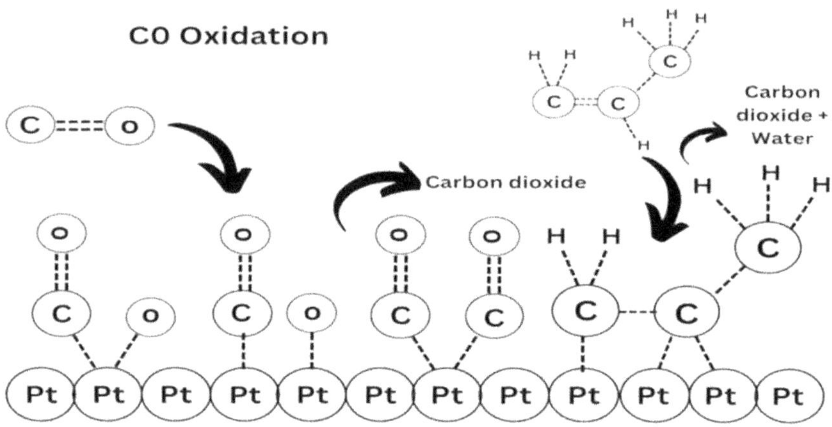


**FIGURE 8.2** The scheme of CO and HC oxidation on Pt catalyst.

PM constitutes a significant component of air pollution and is linked to respiratory and cardiovascular ailments, along with skin disorders.[40] A highly effective method for removing PM from engine exhaust emissions involves capturing it using diesel particulate filters. These DPFs come in various forms, including ceramic and sintered metal-based designs that have been developed for this purpose.[41,42]

Ceria-based catalysts are among the most favorable options for the oxidation of soot within diesel particulate filters. These catalysts possess outstanding redox capabilities and are more cost-effective than noble metals, making them highly attractive choices.[43] Furthermore, the performance and durability of ceria can be enhanced by incorporating metals like rare earth metals, transition metals, alkali metals, and alkaline earth metals into its crystal lattice.[12] Gross and his colleagues.[44] proposed a reaction mechanism for the catalytic combustion of soot on cerium dioxide ($CeO_2$) based on empirical data. This mechanism entails several key steps: oxygen adsorption on the surface of the ceria catalyst, the formation of superoxide and peroxide species, and the reaction of peroxide with soot resulting in the formation of carbon dioxide and the diffusion of oxygen. Figure 8.4 illustrates the schematic representation of this mechanism for soot oxidation using the $CeO_2$–Ag catalyst.[45] A contemporary method to create nanostructured solid solutions of cerium–manganese (CM) and cerium–iron (CF) for soot oxidation involved a straightforward coprecipitation process.[46] A thorough examination of their structural, morphological, and surface characteristics was carried out using various experimental methods, including X-ray diffraction (XRD), inductively coupled plasma optical emission spectroscopy (ICP-OES), BET surface area analysis, scanning electron microscopy with energy-dispersive X-ray spectroscopy (SEM-EDX), and high-resolution transmission electron microscopy (TEM-HRTEM), among others.

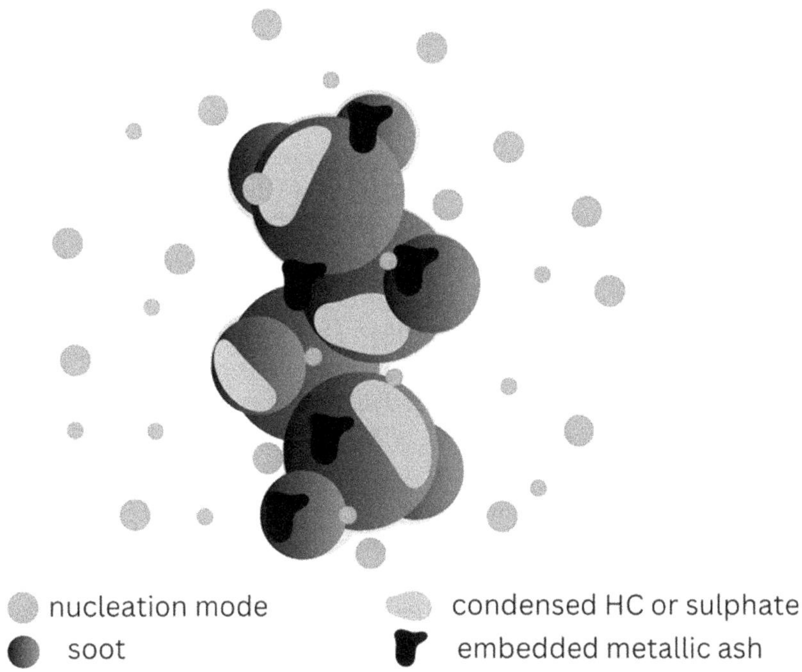


**FIGURE 8.3** Conception of diesel PM.

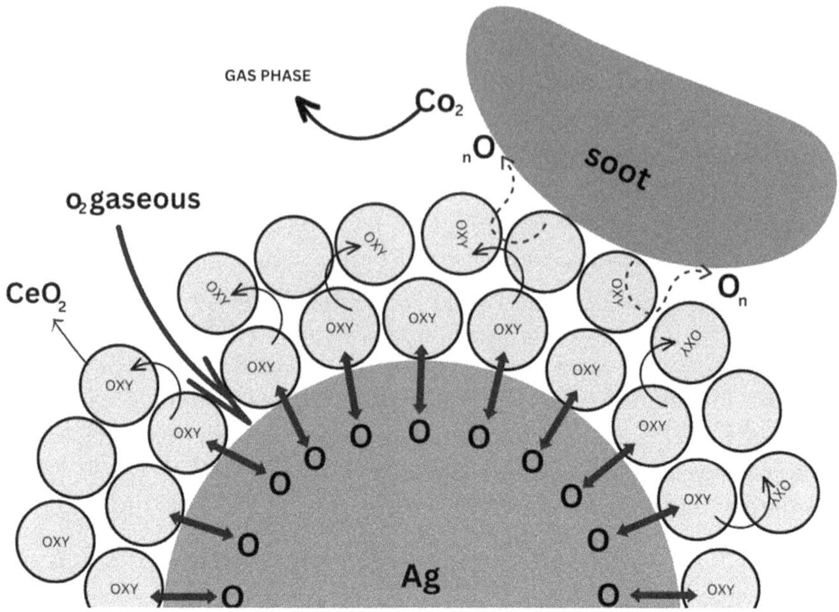


**FIGURE 8.4** Soot oxidation mechanism over the $CeO_2$–Ag catalyst.

The results from XRD and TEM analyses confirmed the creation of nanocrystalline solid solutions where manganese (Mn) and iron (Fe) cations were incorporated into the ceria lattice. SEM studies also revealed that Ce–Mn–O and Ce–Fe–O solid solutions existed in the form of nanoparticles with a uniform distribution. Utilizing UV-visible diffuse reflectance spectroscopy (UV-vis DRS), Raman spectroscopy, and XPS, it was demonstrated that there exists a synergistic interaction between cerium–oxygen (Ce–O) and metal–oxygen (M–O, where M=Manganese or Iron) in the synthesized materials. The catalytic performance of cerium–manganese (CM) and cerium–iron (CF) samples for soot oxidation was evaluated under stringent contact conditions, and comparisons were made with the well-established $CeO_2$–$ZrO_2$ (CZ) catalyst. The doped ceria solid solutions with manganese and iron exhibited enhanced catalytic activity. Furthermore, it was observed that the CM and CF catalysts exhibited greater thermal stability when compared to pure ceria. In terms of catalytic activity, the ranking can be summarized as follows: CM > CF > CZ > C.

## 8.4 THREE-WAY CATALYSTS

Three-way catalyst technology is a commonly employed emission control method in modern vehicles featuring spark ignition engines. It operates by concurrently reducing nitrogen oxides, carbon monoxide, and hydrocarbons within exhaust gases characterized by low oxygen levels, a condition associated with stoichiometric combustion.[9] An essential element of the three-way catalyst system is the catalytic material, which consists of metals from the platinum group. These metals include platinum, palladium, and rhodium among others.[48] This substance serves as a catalyst at active sites, facilitating the reduction of both carbon monoxide and hydrocarbons, as well as the conversion of nitrogen oxides. Noble metals are typically finely distributed throughout ceramic honeycomb structures or monoliths when a highly porous wash coat is applied after impregnation. The wash coat primarily consists of alumina, typically measuring 20 to 40 μm in thickness, with specific surface areas ranging from 100 to 200 $m^2/g$. This composition accounts for approximately 5–15 wt% of the monolith.[10]

In Figure 8.5, you can observe the trifunctional design of the catalyst, serving the purposes of nitrogen oxide reduction and hydrocarbon oxidation. As per this model, achieving effective nitrogen oxide removal requires the presence of three distinct catalytic functions. The first function facilitates the decomposition of NO into $N_2$, while the other two functions enable the oxidation of NO to $NO_2$ and the partial oxidation of hydrocarbons using $NO_2$. The partially oxidized hydrocarbons ultimately undergo complete oxidation when they react with the remaining oxygen generated from the NO decomposition process. Recently, Zeng and Hohn developed a comprehensive and thermodynamically sound surface reaction mechanism to describe the operation of three-way catalysts in treating exhaust emissions from natural gas engines.

Ceria-based catalysts find extensive use in three-way catalyst applications due to ceria's ability to enhance the dispersion of noble metals and its impressive characteristics such as a high surface area and oxygen storage capacity. The incorporation of zirconia into the ceria structure results in the creation of a substantial number of defects, leading to a significant enhancement in ceria's thermal stability and oxygen storage capacity.[51] During the experimental investigations involving a range of catalytic samples, researchers monitored and analyzed the changes in components including CO, $C_3H_6$, $O_2$, and NO. Additionally, the impact of prereduction and aging treatments in cyclic redox feed streams was examined at a temperature of 1173 K under various conditions. The enhancement of $CeO_2$–$ZrO_2$ catalysts can be achieved through their modification, which involves the introduction of different metals into the crystal lattice. Li et al. conducted a study on Pd/$CeO_2$–$ZrO_2$ catalysts that had been modified with various transition metals such as Cr, Mn, Ni, Co, and Fe.[52]

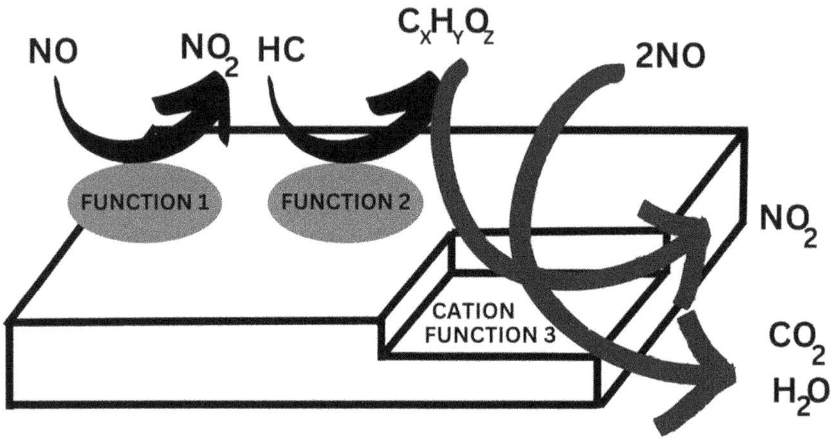


**FIGURE 8.5** The three-way functional model for designing $DeNO_x$ and HC oxidation catalysts.

Catalysts containing rhodium demonstrate the highest level of efficiency when it comes to converting NO.[53,54] Utilizing $ZrO_2$ as a support material significantly improves both the catalytic performance and long-term durability of rhodium catalysts, whether in powder or monolith form, when subjected to diverse aging conditions.[53] The rhodium particles in Rh/$ZrO_2$ were notably smaller in size compared to those in Rh/$Al_2O_3$, particularly following exposure to air and cyclic aging. $ZrO_2$ also exhibited the capacity to retain a higher quantity of surface rhodium species and sustain the surface state of rhodium in a relatively reduced state compared to $Al_2O_3$ support. These findings confirm the advantages and suitability of $ZrO_2$ as an effective catalyst support for rhodium. To assess the performance of honeycomb-coated rhodium catalysts in three-way catalysis, various metal phosphate supports were examined under different simulated exhaust conditions, involving a range of air-to-fuel ratios.[54]

These catalysts' support significantly enhanced the efficiency of nitrogen oxide conversion under slightly lean conditions (with air-to-fuel ratios falling between 14.6 and 15.3) when compared to conventional $Rh/ZrO_2$ catalysts. The denitrification activity followed a sequence of increasing mean electronegativity among the supports: $ZrO_2$ < $YPO_4$ < $LaPO_4$ < $AlPO_4$ < $ZrP_2O_7$. While both NO–Co and NO–$C_3H_6$ reactions contributed to the denitrification process, the latter reaction was more favored on $Rh/ZrP_2O_7$ compared to $Rh/ZrO_2$.

## 8.5 SELECTIVE CATALYTIC REDUCTION (SCR)

While three-way catalysts effectively decrease nitrogen oxide emissions in gasoline engines with low oxygen levels, they are not suitable for use in lean-burn gasoline and diesel engines when oxygen is present.[55] Within the realm of technologies aimed at reducing nitrogen oxide emissions in lean conditions, the SCR technology stands out for its notable effectiveness, cost-effectiveness, long-lasting performance, and the presence of readily available infrastructure.[15] In urea-based SCR, urea is stored on board and introduced into the exhaust system before the SCR catalyst. It is then converted into ammonia, which interacts with nitrogen oxides on the catalyst's surface, ultimately reducing them to nitrogen ($N_2$).

The widely recognized and commercially used catalyst in automotive applications is composed of vanadium dispersed on a titanium support, with the added promotion of tungsten trioxide ($V_2O_5$–$WO_3/TiO_2$).[56] This commercially available catalyst exhibits significant activity and selective reduction of nitrogen oxides to nitrogen ($N_2$) in $NH_3$ SCR at temperatures ranging from 300 to 400°C. Mn-based metal oxides, on the other hand, demonstrate effective low-temperature SCR performance.[11] Hu et al. put forth a potential reaction mechanism for the Co–Mn/$TiO_2$ catalyst.[57] In this proposed mechanism, the Co–Mn/$TiO_2$ catalyst initially adsorbs NO molecules onto its surface, resulting in the formation of four distinct nitrogen oxide species: gaseous $NO_2$, bidentate nitrates, linear nitrites, and monodentate nitrites. Meanwhile, the $NH_3$ reactant gas is adsorbed onto Lewis acid sites in a coordinated $NH_3$ form. At lower temperatures, only gaseous $NO_2$ actively participates in the catalytic reaction, serving as a rapid intermediate in the SCR process. However, at elevated temperatures, all of these nitrogen oxide species readily engage with the coordinated $NH_3$, leading to the production of $N_2$.

Two variants of the $V_2O_5$–$WO_3/TiO_2$ catalyst were synthesized using different techniques, including coprecipitation and incipient wetness impregnation methods.[58] The catalytic activities of SCR were thoroughly studied, and the distinctions and similarities between two preparation methods were identified. Both coprecipitation and impregnation methods led to the formation of surface mono-oxo sites, namely, $O{=}VO_3$ and $O{=}WO_4$, on the $TiO_2$ support for VTi, WTi, and VWTi catalysts. However, it is worth noting that the coprecipitated catalysts containing V and W introduced new surface $O{=}VO_3$ and $O{=}WO_4$ sites on the $TiO_2$, in contrast to simply anchoring them to existing surface defects. Consequently, the SCR activity of the coprecipitated catalysts consistently outperformed the impregnated ones, even when they had the same catalyst composition and surface area. The enhanced performance of coprecipitated VWTi catalysts compared to the impregnated ones can be attributed to their greater capacity for ammonia adsorption. This ability is closely linked to the higher presence of surface defects on the $TiO_2$ support in the coprecipitated catalysts, relative to the lower defect concentration on the highly crystalline $TiO_2$ support in the impregnated catalysts.

SCR systems have evolved through three commercial generations and are now achieving a remarkable nitrogen oxide removal efficiency of approximately 95 to 96%.[9] However, continuous advancements are underway in the enhancement of these systems. Focus areas include catalyst development, durability, management of ammonia adsorption, and system design. Notably, Zhao et al. introduced a novel catalyst featuring zirconium-doped $CeVO_4$ for low-temperature SCR of nitrogen oxides using $NH_3$, effectively covering a temperature range from 150 to 375°C.[59]

The process schematic, depicted in Figure 8.6, illustrates the impact of zirconium (Zr) doping into $CeVO_4$. This doping results in an augmentation of active oxygen species and oxygen vacancies. Simultaneously, it elevates the presence of both Bronsted acid sites and Lewis acid sites, all of which contribute positively to the $NH_3$–SCR reactions. Cheng and colleagues conducted research to explore the enhancement of ceria's promotional effect on SCR activity compared to $V_2O_5$–$WO_3/TiO_2$.[60] Through a homogeneous precipitation method, composite supports consisting of $CeO_2$ and $TiO_2$ were developed with varying Ce/Ti molar ratios. Subsequently, $V_2O_5$–$WO_3/TiO_2$ catalysts for the selective catalytic reduction of nitrogen oxides using $NH_3$ were prepared via an incipient wetness impregnation technique. During the experimental investigations, it became evident that the catalytic activity

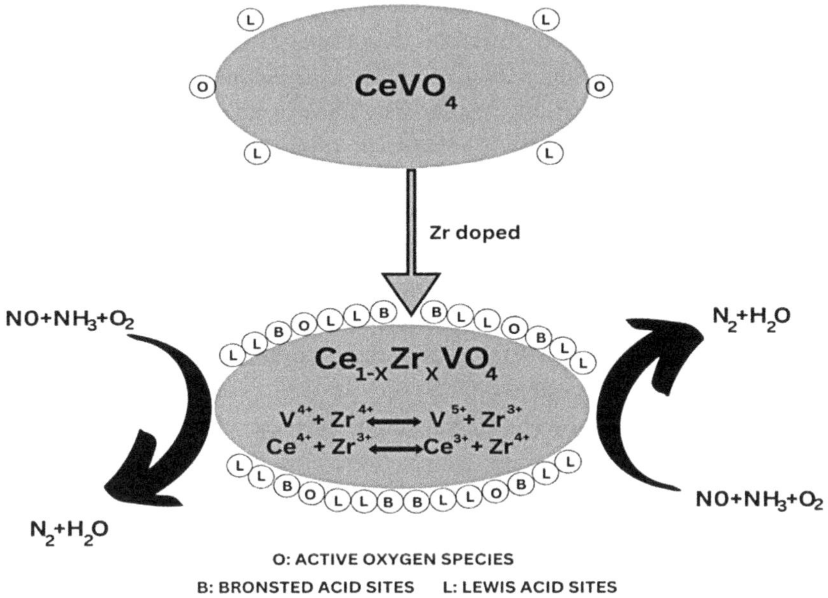


**FIGURE 8.6** The relationships of active oxygen species, acid sites, and catalytic performance.

of $V_2O_5$–$WO_3$/$TiO_2$ was significantly enhanced through the incorporation of Ce into the $TiO_2$ support. This improvement was attributed to Ce's ability to augment oxygen adsorption, a crucial factor in nitrogen oxide reduction. Additionally, Song and colleagues explored another positive aspect of Ce doping in $V_2O_5$–$WO_3$/$TiO_2$ catalysts, specifically its impact on SCR activity.[61] As is commonly recognized, sulfur is present in all fuels, consequently resulting in the presence of $SO_2$ in exhaust gases. In a steam-rich atmosphere, $SO_2$ reacts with $NH_3$, giving rise to the formation of ammonia bisulfate (ABS). ABS tends to accumulate on the active sites of the catalyst surface, particularly at lower temperatures (below 220°C, which is the dew point of ABS). This accumulation leads to the deactivation of SCR catalysts. To address this issue, researchers conducted a study involving the decomposition of ABS over $CeO_2$ and $V_2O_5$–$WO_3$/$CeO_2$–$TiO_2$ catalysts, employing TG-MS (Thermogravimetric Analysis-Mass Spectrometry) analysis. The results demonstrated that $CeO_2$ facilitates the decomposition of ABS, leading to the rejuvenation of the catalyst. Furthermore, it is worth noting that metal-exchanged zeolite SCR catalysts, such as CuZ and FeZ, are widely employed in passenger and light-duty diesel vehicles.[62]

Vennestrom and colleagues conducted a study in which they explored the effectiveness of two different copper-exchanged zeolites, namely, ZSM-5 and IM-5, for the selective catalytic reduction of nitrogen oxides using ammonia.[63]. Initially, both Cu–ZSM-5 and Cu–IM-5 zeolites exhibit similar catalytic activity. However, after undergoing hydrothermal aging, Cu–IM-5 experiences significantly less deactivation compared to Cu–ZSM-5. This difference in deactivation is primarily attributed to the superior framework stability of Cu–IM-5. SCR catalysts were developed by combining Fe–ZSM-5 zeolite with mechanical mixtures containing Mn oxides and binary Mn–X (where X can be Ce, Cr, Cu, or Zr ) mixed oxides. The combination of these materials results in potent and, at times, substantial synergistic effects.[64]

Feng and colleagues conducted research to examine the influence of cerium (Ce) as a promoter on a CuFe/beta monolith catalyst used in the selective catalytic reduction of nitrogen oxides with ammonia.[65] Both $Cu_3Fe_3$/beta and $Cu_3Fe_3$/Ce/beta catalysts were synthesized using the incipient wetness impregnation method, followed by a comprehensive investigation of their selective catalytic reduction activity and resistance to hydrothermal conditions. The experimental findings highlighted the superior $NH_3$–SCR performance of $Cu_3Fe_3$/Ce/beta catalysts, attributed to the enhanced dispersion of Fe and/or Cu species within this catalyst compared to $Cu_3Fe_3$/beta. Furthermore, the addition of Ce demonstrated an additional benefit by improving the hydrothermal stability of $Cu_3Fe_3$/beta. Additionally, other SCR catalysts based on mesoporous silica, specifically MCM-41, were also subject to investigation.[66]

MCM-41, a mesoporous silica material, underwent doping with various metal oxides like Ru, Co, Ni, Fe, and Cu using a coprecipitation method. These doped catalysts were then subjected to testing in the selective catalytic reduction (SCR) of nitric oxide (NO) with carbon monoxide (CO) across a temperature range spanning from 223 to 723 Kelvin. Under reaction conditions below 550 Kelvin, the catalysts exhibited efficiencies in the following order: Ru–MCM-41 > Co–MCM-41 > Ni–MCM-41 > Fe–MCM-41 > Cu–MCM-41. However, when operating at temperatures exceeding 650 Kelvin, Ru–MCM-41 emerged as the most active catalyst, with Cu–MCM-41 following closely behind. Furthermore, the study investigated three cerium-doped activated carbon materials, namely, activated carbon fiber (ACF) and multiwalled carbon nanotubes (MWCNTs), as potential catalysts for reducing NO emissions at the operating temperatures typical of diesel engines.[67]

Jeong and colleagues conducted research to explore the synergistic impact of catalysts supported on mixed Mg–Al oxides in comparison to those supported on γ-$Al_2O_3$.[68] In the temperature range of 250 to 500°C, it was observed that when Pt and Ba were loaded onto catalysts, the Mg–Al mixed oxide support outperformed the γ-$Al_2O_3$-supported catalyst in terms of nitrogen oxide storage capacity. The optimal Mg/Al ratio fell within the range of 2:8 to 4:6. Additionally, the presence of Mg in thermally aged catalysts resulted in several positive effects, including enhanced thermal stability by mitigating Pt sintering and reducing the formation of the crystalline $BaAl_2O_4$ phase. Among the catalysts with varying Mg/Al ratios, the Pt–BaO/Mg–Al mixed oxide support with a Mg–Al ratio of 4:6 exhibited the highest nitrogen oxide storage capacity following thermal treatment at 800°C.[69]

## 8.6 CONCLUSIONS

In conclusion, this comprehensive examination has provided insights into various catalytic technologies employed for controlling vehicle exhaust emissions. With a primary focus on diesel engines, the discussion covered the critical components of emissions control, including diesel oxidation catalysts, diesel particulate filters, three-way catalysts, and selective catalytic reduction systems. Diesel oxidation catalysts play a vital role in converting harmful emissions such as carbon monoxide, hydrocarbons, and nitrogen oxides into less harmful compounds like water and carbon dioxide. Recent advancements have explored the use of various catalyst compositions and support materials to enhance their performance and durability. Diesel particulate filters are crucial for capturing soot particles present in diesel exhaust emissions. Innovative catalysts, particularly cerium-based catalysts, have shown great promise in efficiently oxidizing soot and improving the durability of DPFs.

Three-way catalysts are widely employed in gasoline engines to simultaneously reduce nitrogen oxides, carbon monoxide, and hydrocarbons under stoichiometric conditions. These catalysts typically contain noble metals like platinum, palladium, and rhodium, and their performance is continually optimized. Selective catalytic reduction systems are essential for reducing nitrogen oxide emissions in lean-burn engines, particularly diesel engines. Various catalyst compositions, including vanadium-based and cerium-based catalysts, have been explored to achieve efficient $NO_x$ reduction using ammonia as a reducing agent. The incorporation of nano-technology into catalytic systems offers exciting prospects for improving catalyst efficiency and reducing precious metal usage. Nanostructured catalysts have demonstrated the ability to enhance catalytic activity and stability, making them a promising avenue for future research and development. Advancements in catalytic technologies continue to play a crucial role in mitigating the environmental and health impacts of vehicle exhaust emissions. These innovations enable the automotive industry to meet stringent emission standards while striving for greater efficiency and sustainability in transportation.

## KEYWORDS

- **nanoparticles**
- **vehicle exhaust**
- **nanostructured catalyst**
- **emission control**
- **green technology**
- **environmental sustainability**

## REFERENCES

1. Bielaczyc, P.; Woodburn, J.; Szczotka, A. An Assessment of Regulated Emissions and $CO_2$ Emissions from a European Light-Duty CNG-Fueled Vehicle in the Context of Euro 6 Emissions Regulations. *Appl. Energy* **2014,** *117*, 134–141.
2. Graham, L. A.; Rideout, G.; Rosenblatt, D.; Hendren, J. Greenhouse Gas Emissions from Heavy-Duty Vehicles. *Atmos. Environ.* **2008,** *42*, 4665–4681.
3. Yan, F.; Winijkul, E.; Jung, S.; Bond, T. C.; Streets, D. G. Global Emission Projections Of Particulate Matter (PM): I. Exhaust Emissions from On-Road Vehicles. *Atmos. Environ.* **2011,** *45*, 4830–4844.
4. Pope, C. A.; Burnett, R. T.; Thun, M. J.; Calle, E. E.; Krewski, D.; Ito, K.; Thurston, G. D. Lung Cancer, Cardiopulmonary Mortalityand Long-Term Exposure to Fine Particulate Air Pollution. *J. Am. Med. Assoc.* **2002,** *287*, 1132–1141.
5. Speight, J. G. *The Chemistry and Technology of Petroleum*; CRC Press: New York, 2014.

6. Li, C.; Zhao, X.; Wang, A.; Huber, G. W.; Zhang, T. Catalytic Transformation of Lignin for the Production of Chemicals and Fuels. *Chem. Rev.* **2015,** *115*, 11559–11624.
7. Johnson, T. Review of Vehicular Emissions Trends. SAE Technical Paper 2015–01–0993; 2015.
8. Gerasimov, G. Nanomaterials in Proton Exchange Fuel Cells. *J. Eng. Phys. Thermophys.* **2015,** *88*, 1554–1568.
9. Johnson, T. Vehicular Emissions in Review. SAE Technical Paper 2013–01–0538; 2013.
10. Granger, P.; Parvulescu, V. I. Catalytic NOx Abatement Systems for Mobile Sources: From Three-Way to Lean Burn After-Treatment Technologies. *Chem. Rev.* **2011,** *111*, 3155–3207.
11. Guan, B.; Zhan, R.; Lin, H.; Huang, Z. Review of the State-of-the-Art of Exhaust Particulate Filter Technology in Internal Combustion Engines. *J. Environ. Manag.* **2015,** *154*, 225–258.
12. Fino, D.; Bensaid, S.; Piumetti, P.; Russo, N. A Review on the Catalytic Combustion of Soot in Diesel Particulate Filters for Automotive Applications: From Powder Catalysts to Structured Reactors. *Appl. Catal. A* **2016,** *509*, 75–96.
13. Russell, A.; Epling, W. S. Diesel Oxidation Catalysts. *Catal. Rev.* 2011; *53*, 337–423.
14. Prasad, R.; Bella, V. R. A Review on Diesel Soot Emission, Its Effect and Control. *Bull. Chem. React. Eng. Catal.* **2010,** *5*, 69–87.
15. Guan, B.; Zhan, R.; Lin, H.; Huang, Z. Review of State of the Art Technologies of Selective Catalytic Reduction of NOx from Diesel Engine Exhaust. *Appl. Thermal Eng.* **2014,** *66*, 395–414.
16. Roy, S.; Baiker, A. NOx Storage-Reduction Catalysis: From Mechanism and Materials Properties to Storage-Reduction Performance. *Chem. Rev.* **2009,** *109*, 4054–4091.
17. Liu, G.; Gao, P.-X. A Review of NOx Storage/Reduction Catalysts: Mechanism, Materials and Degradation Studies. *Catal. Sci. Technol.* **2011,** *1*, 552–568.
18. Pitkethly, M. J. Nanomaterials: The Driving Force. *Mater. Today* **2004,** *7* (12), 20–29.
19. Liu, C.-J.; Burghaus, U.; Besenbacher, F.; Wang, Z. L. Preparation and Characterization of Nanomaterials for Sustainable Energy Production. *ACS Nano* **2010,** *4*, 5517–5526.
20. Kharisov, B. I.; Kharissovaand, O. V.; Ortiz-Mendez, U. (Eds.) *CRC Concise Encyclopedia of Nanotechnology*; CRC Press, Taylor & Francis Group: New York, 2015.
21. Chaturvedi, S.; Dave, P. N.; Shah, N. K. Applications of Nano-Catalyst in New Era. *J. Saudi Chem. Soc.* **2012,** *16*, 307–325.
22. Gololobov, A. M.; Bekk, I. E.; Bragina, G. O.; Zaikovskii, V. I.; Ayupov, A. B.; Telegina, N. S.; Bukhtiyarov, V. I.; Stakheev, A. Yu. Platinum Nanoparticle Size Effect on Specific Catalytic Activity in n-Alkane Deep Oxidation: Dependence on the Chain Length of the Paraffin. *Kinet. Catal.* **2009,** *50*, 830–836.
23. Behafarid, F.; Ono, L. K.; Mostafa, S.; Croy, J. R.; Shafai, G.; Hong, S.; Rahman, T. S.; Bare, S. R.; Cuenya, B. R. Electronic Properties and Charge Transfer Phenomena in Pt Nanoparticles on $\gamma$-$Al_2O_3$: Size, Shape, Support and Adsorbate Effects. *Phys. Chem. Chem. Phys.* **2012,** *14*, 11766–11779.
24. Li, L.; Larsen, A. H.; Romero, N. A.; Morozov, V. A.; Glinsvad, C.; Abild-Pedersen, F.; Greeley, J.; Jacobsen, K. W.; Nørskov, J. K. Investigation of Catalytic Finite-Size-Effects of Platinum Metal Clusters. *J. Phys. Chem. Lett.* **2013,** *4*, 222–226.
25. Cuenya, B. R.; Behafarid, F. Nanocatalysis: Size- and Shape-Dependent Chemisorption and Catalytic Reactivity. *Surf. Sci. Rep.* **2015,** *70*, 135–187.

26. Lizarraga, L.; Souentie, S.; Boreave, A.; George, C.; D'Anna, B.; Vernoux, P. Effect of Diesel Oxidation Catalysts on the Diesel Particulate Filter Regeneration Process. *Environ. Sci. Technol.* **2011,** *45*, 10591–10597.
27. Kim, C.; Schmid, M.; Schmieg, S.; Tan, J.; Li, W. The Effect of Pt-Pd Ratio on Oxidation Catalysts Under Simulated Diesel Exhaust. SAE Technical Paper 2011–01–1134; 2011.
28. Xiong, H.; Peterson, E.; Qi, G.; Datye, A. K. Trapping Mobile Pt Species by PdO in Diesel Oxidation Catalysts: Smaller Is Better. *Catal. Today* **2016,** *272*, 80–86.
29. Patterson, M. J.; Angove, D. E.; Cant, N. W. The Effect of Carbon Monoxide on the Oxidation of Four C6 to C8 Hydrocarbons Over Platinum, Palladium and Rhodium. *Appl. Catal. B* **2000,** *26*, 47–57.
30. Salomons, S.; Votsmeier, M.; Hayes, R.; Drochner, A.; Vogel, H.; Gieshof, J. CO and $H_2$ Oxidation on a Platinum Monolith Diesel Oxidation Catalyst. *Catal. Today* **2006,** *117*, 491–497.
31. Hazlett, M. J.; Moses-Debusk, M.; Parks, J. E. II, Allard, L. F.; Epling, W. S. Kinetic and Mechanistic Study of Bimetallic Pt-Pd/$Al_2O_3$ Catalysts for CO and $C_3H_6$ Oxidation. *Appl. Catal. B* **2017,** *202*, 404–417.
32. Smeltz, A.; Getman, R.; Schneider, W.; Ribeiro, F. Coupled Theoretical and Experimental Analysis of Surface Coverage Effects in Pt-Catalyzed NO and $O_2$ Reaction to $NO_2$ on Pt(111). *Catal. Today* **2008,** *136*, 84–92.
33. Fayad, M. A.; Herreros, J. M.; Martos, F. J.; Tsolakis, A. Role of Alternative Fuels on Particulate Matter (PM) Characteristics and Influence of the Diesel Oxidation Catalyst. *Environ. Sci. Technol.* **2015,** *49*, 11967–11973.
34. Wong, A. P.; Kyriakidou, E. A.; Toops, T. J.; Regalbuto, J. R. The Catalytic Behavior of Precisely Synthesized Pt–Pd Bimetallic Catalysts for Use as Diesel Oxidation Catalysts. *Catal. Today* **2016,** *267*, 145–156.
35. Huang, H.; Jiang, B.; Gu, L.; Qi, Z.; Lu, H. Promoting Effect of Vanadium on Catalytic Activity of Pt/Ce–Zr–O Diesel Oxidation Catalysts. *J. Environ. Sci.* **2015,** *33*, 135–142.
36. Piumetti, M.; Bensaid, S.; Fino, D.; Russo, N. Nanostructured Ceria-Zirconia Catalysts for CO Oxidation: Study on Surface Properties and Reactivity. *Appl. Catal. B* **2016,** *197*, 35–46.
37. Kim, M.-Y.; Kyriakidou, E. A.; Choi, J.-S.; Toops, T. J.; Binder, A. J.; Thomas, C.; Parks, J. E. II, Schwartz, V.; Chen, J.; Hensley, D. K. Enhancing Low-Temperature Activity and Durability of Pd-Based Diesel Oxidation Catalysts Using $ZrO_2$ Supports. *Appl. Catal. B* **2016,** *187*, 181–194.
38. Sun, X.; Su, H.; Lin, Q.; Han, C.; Zheng, Y.; Sun, L.; Qi, C. Au/Cu–Fe–La–$Al_2O_3$: A Highly Active, Selective and Stable Catalysts for Preferential Oxidation of Carbon Monoxide. *Appl. Catal. A* **2016,** *527*, 19–29.
39. Maricq, M. M. Chemical Characterization of Particulate Emissions from Diesel Engines: A Review. *J. Aerosol Sci.* **2007,** *38*, 1079–1118.
40. Frank, B.; Schlögl, R.; Su, D. S. Diesel Soot Toxification. *Environ. Sci. Technol.* **2013,** *47*, 3026–3027.
41. Mathur, S.; Johnson, J. H.; Naber, J.; Bagley, S. T.; Shende, A. S. Experimental Studies of an Advanced Ceramic Diesel Particulate Filter. SAE Technical Paper 2008–01–0622; 2008.

42. Kamp, C.; Folino, P.; Wang, Y.; Sappok, A.; Ernstmeyer, J.; Saeid, A.; Singh, R.; Kharraja, B.; Wong, V. Ash Accumulation and Impact on Sintered Metal Fiber Diesel Particulate Filters. SAE Technical Paper 2015–01–1012; 2015.
43. Bueno-López, A. Diesel Soot Combustion Ceria Catalysts. *Appl. Catal. B* **2014,** *146*, 1–11.
44. Gross, M. S.; Ulla, M. A.; Querini, C. A. Diesel Particulate Matter Combustion with $CeO_2$ as Catalyst. Part I: System Characterization and Reaction Mechanism. *J. Mol. Catal. A* **2012,** *352*, 86–94.
45. Yamazaki, K.; Kayama, T.; Dong, F.; Shinjoh, H. A Mechanistic Study on Soot Oxidation Over $CeO_2$–Ag Catalyst with 'Rice-Ball' Morphology. *J. Catal.* **2011,** *282*, 289–298.
46. Venkataswamy, P.; Jampaiah, D.; Rao, K. N.; Reddy, B. M. Nanostructured Ce0.7Mn0.3O2−δ and $Ce_{0.7}Fe_{0.3}O_{2-\delta}$ Solid Solutions for Diesel Soot Oxidation. *Appl. Catal. A* **2014,** *488*, 1–10.
47. Piumetti, M.; Bensaid, S.; Russo, N.; Fino, D. Investigations into Nanostructured Ceria–Zirconia Catalysts for Soot Combustion. *Appl. Catal. B* **2016,** *180*, 271–282.
48. Matsuzono, Y.; Kuroki, K.; Nishi, T.; Suzuki, N.; Yamada, T.; Hirota, T.; Zhang, Z. Development of Advanced and Low PGM TWC System for LEV2 PZ EV and LEV3 SULEV30. SAE Technical Paper 2012–01–1242; 2012.
49. Djéga-Mariadassou, G.; Boudart, M. Classical Kinetics of Catalytic Reactions. *J. Catal.* **2003,** *216*, 89–97.
50. Zeng, F.; Hohn, K. L. Modeling of Three-Way Catalytic Converter Performance with Exhaust Mixture from Natural Gas-Fueled Engines. *Appl. Catal. B* **2016,** *182*, 570–579.
51. Gonzalez-Marcos, M. P.; Pereda-Ayo, B.; Aranzabal, A.; Gonzalez-Marcos, J. A.; Gonzalez-Velasco, J. R. On the Effect of Reduction and Ageing on the TWC Activity of Pd/Ce0.68Zr0.32O2 Under Simulated Automotive Exhausts. *Catal. Today* **2012,** *180*, 88–95.
52. Li, G.; Wang, Q.; Zhao, B.; Zhou, R. A New Insight Into the Role of Transition Metals Doping with $CeO_2$–$ZrO_2$ and Its Application in Pd-Only Three-Way Catalysts for Automotive Emission Control. *Fuel* **2012,** *92*, 360–368.
53. Cao, Y.; Ran, R.; Wu, X.; Zhao, B.; Weng, D. Improved Activity and Durability of Rh-Based Three-Way Catalyst Under Diverse Aging Atmospheres by $ZrO_2$ Support. *J. Environ. Sci.* **2016**. http://dx.doi.org/10.1016/j.jes.2016.04.017
54. Nagao, Y.; Nakahara, Y.; Sato, T.; Iwakura, H.; Takeshita, S.; Minami, S.; Yoshida, H.; Machida, M. Rh/ZrP2O7 as an Efficient Automotive Catalyst for NOx Reduction Under Slightly Lean Conditions. *ACS Catal.* **2015,** *5*, 1986–1994.
55. Li, J.; Chang, H.; Ma, L.; Hao, J.; Yang, R. T. Low-Temperature Selective Catalytic Reduction of NOx with $NH_3$ Over Metal Oxide and Zeplitecaralyst: A Review. *Catal. Today* **2011,** *175*, 147–156.
56. Putluru, S. S. R.; Schill, L.; Godiksen, A.; Poreddy, R.; Mossin, S.; Jensen, A. D.; Fehrmann, R. Promoted $V_2O_5/TiO_2$ Catalysts for Selective Catalytic Reduction of NO with $NH_3$ at Low Temperatures. *Appl. Catal. B* **2016,** *183*, 282–290.
57. Hu, H.; Cai, S.; Li, H.; Huang, L.; Shi, L.; Zhang, D. Mechanistic Aspects of deNOx Processing Over $TiO_2$ Supported Co-Mn Oxide Catalysts: Structure–Activity Relationships and in situ DRIFTs Analysis. *ACS Catal.* **2015,** *5*, 6069–6077.

58. He, Y.; Ford, M. E.; Zhu, M.; Liu, Q.; Tumuluri, U.; Wu, Z.; Wachs, I. E. Influence of Catalyst Synthesis Method on Selective Catalytic Reduction (SCR) of NO by $NH_3$ with $V_2O_5$-$WO_3$/$TiO_2$ Catalysts. *Appl. Catal. B* **2016,** *193*, 141–150.
59. Zhao, X.; Huang, L.; Li, H.; Hu, H.; Hu, X.; Shi, L.; Zhang, D. Promotional Effects of Zirconium Doped CeVO4 for the Low-Temperature Selective Catalytic Reduction of NOx with $NH_3$. *Appl. Catal. B* **2016,** *183*, 269–281.
60. Cheng, K.; Liu, J.; Zhang, T.; Li, J.; Zhao, 2016Z.; Wei, Y.; Jiang, G.; Duan, A. Effect of Ce Doping of $TiO_2$ Support on NH3-SCR Activity Over $V_2O_5$–$WO_3$/$CeO_2$–$TiO_2$ Catalyst. *J. Environ. Sci.* **2014,** *26*, 2106–2113.
61. Song, L.; Chao, J.; Fang, Y.; He, H.; Li, J.; Qiu, W.; Zhang, G. Promotion of Ceria for Decomposition of Ammonia Bisulfate Over $V_2O_5$-$MoO_3$/$TiO_2$ Catalyst for Selective Catalytic Reduction. *Chem. Eng. J.* **2016,** *303*, 275–281.
62. Pereira, M. V. L.; Nicolle, A.; Berthout, D. Hydrothermal Aging Effects on Cu-zeolite $NH_3$-SCR Catalyst. *Catal. Today* **2015,** *258*, 424–431.
63. Vennestrom, P. N. R.; Janssens, T. V. W.; Kustov, A.; Grill, M.; Puig-Molina, A.; Lundegaard, L. F.; Tiruvalam, R. R.; Concepcion, P.; Corma, A. Influence of Lattice Stability on Hydrothermal Deactivation of Cu-ZSM-5 and Cu-IM-5 Zeolites for Selective Catalytic Reduction of NOx by $NH_3$. *J. Catal.* **2014,** *309*, 477–490.
64. Salazar, M.; Becker, R.; Grünert, W. Hybrid Catalysts: An Innovative Route to Improve Catalyst Performance in the Selective Catalytic Reduction of NO by $NH_3$. *Appl. Catal. B* **2015,** *165*, 316–327.
65. Kwak, J. H.; Tonkyn, R. G.; Kim, D. H.; Szanyi, J.; Peden, C. H. F. Excellent Activity and Selectivity of Cu-SSZ-13 in the Selective Catalytic Reduction of NOx with NH3. *J. Catal.* **2010,** *275*, 187–190.
66. Feng, X.; Cao, Y.; Lan, L.; Lin, C.; Li, Y.; Xu, H.; Gong, M.; Chen, Y. The Promotional Effect of Ce on CuFe/Beta Monolith Catalyst for Selective Catalytic Reduction of NOx by Ammonia. *Chem. Eng. J.* **2016,** *302*, 697–706.
67. Patel, A.; Shukla, P.; Rufford, T. E.; Rudolph, V.; Zhu, Z. Selective Catalytic Reduction of NO with CO Using Different Metal-Oxides Incorporated in MCM-41. *Chem. Eng. J.* **2014,** *255*, 437–444.
68. Athappan, A.; Sattler, M. L.; Sethupathi, S. Selective Catalytic Reduction of Nitric Oxide Over Cerium-Doped Activated Carbons. *J. Environ. Chem. Eng.* **2015,** *3*, 2502–2513.
69. Jeong, S.; Youn, S.; Kim, D. H. Effect of Mg/Al Ratios on the NOx Storage Activity Over Pt-BaO/Mg–Al Mixed Oxides. *Catal. Today* **2014,** *231*, 155–163.

## CHAPTER 9

# Sustainable Algal Fuel Cells: Paving the Way for Green Energy in Space Missions

EASWAR VIVEK ANAND, POOJA SHAJI, and SOOSEN SAMUEL M.

*Department of Physics, Catholicate College, Pathanamthitta, Kerala, India*

## ABSTRACT

Exploring sustainable and reliable energy sources for extended space missions and interplanetary colonization is an inevitable necessity. Algal fuel cells have emerged as a promising technology, providing renewable green energy suitable for consideration in space exploration. This study aims to evaluate the feasibility and advantages of integrating algal fuel cells into space systems. The photosynthetic capability of algae, converting light energy into electrical energy through biochemical processes, offers a convenient method for generating electricity in algal fuel cells. Ongoing research on algae in space indicates their resilience in harsh environments, displaying robust growth while maintaining essential physiological processes. A significant challenge that hinders the advancement of algal fuel cells is their lower efficiency. Certainly, these challenges can be addressed through various strategies, such as refining and adapting electrodes, identifying optimal growth conditions for algae, and altering the characteristics of algae varieties. Extensive research on biofuel cells has demonstrated their feasibility, and advancements continue to bolster the case for considering algal fuel cells (AFCs) in space system applications. Algae's inherent ability to generate oxygen and edible food renders them

Polymers and Functional Materials for a Cleaner Environment and Human Health: Sustainable Technologies and New Research. Ruby Varghese, Ann Rose Abraham, Shrikaant Kulkarni, & A. K. Haghi (Eds.)

a standout choice for life support systems in space. Hybrid systems can be engineered utilizing algae, serving the dual purpose of wastewater treatment and electricity generation simultaneously. The utilization of algae-based systems with these distinctive features has proven highly beneficial for space missions. The integration of AFCs into space systems not only holds promise for sustainable energy solutions but also reinforces the viability of algae-based technologies for the future of space exploration.

## 9.1 INTRODUCTION

Fossil fuels are being used as a major source of energy production. However, the depletion of fossil fuels and increased amount of carbon dioxide emission[1] has escalated the necessity of using renewable energy sources. They have been of major importance in producing electricity because of their remarkable properties and green nature. Replacing conventional fossil fuel-based energy production systems with renewable systems has become substantial lately.

The capabilities of renewable energy sources can be very advantageous for energy production not only in our environment but also for outer earth space explorations. The reduced emission of greenhouse gases and sustainability over a long time with reliable energy security makes them ideal for this. However, the choices for the source of energy are very limited in space. Solar energy has been the only renewable energy source available in space that is used for energy production.[2] However, introducing biomaterials and plants to space and producing bioenergy can change this. Biofuel cells can be a major source of renewable energy in space. The characteristics of biofuel cells exhibit variations based on the chosen materials. Optimal performance, efficiency, and convenience are essential criteria in material selection. Algae emerge as a pivotal biomaterial due to its exceptional properties, making it a highly suitable candidate for productive utilization. Consequently, the exploration of algal fuel cells for energy production in space presents significant opportunities and holds substantial promise.

Recent advances in microbial fuel cells have added a great scope for bioelectricity generation. AFCs represent a promising, novel, cost-effective, environment-benign technology for sustainable energy production.[3] They also acquire notable interest, especially due to their potential as a renewable energy source in conjunction with wastewater treatment,[4] which is another mentionable feature useful in space. Since algae uses photosynthesis to produce energy, it results in the production of oxygen[5] and if this system is integrated into life support in space, this oxygen can be used for that purpose also.[6] Carbon dioxide removal is a separate application that can also be incorporated into this. Cultivating microalgae in photobioreactors (PBR) allows them to be used for $CO_2$ fixation under a variety of conditions. In photobioreactors, photosynthesis takes place in which microalgae convert incident light and $CO_2$ into biomass.[7] The incorporation of algae could account for various needs in space exploration, including food to waste treatment and energy production.

Algal fuel cells are an emerging technology where the photosynthetic capability of algae is utilized for producing electricity. Their ability to harness solar energy and biochemically convert it into electrical energy makes it an excellent renewable energy source which is green in nature. Algae are recognized as a valuable nutrient source for space exploration, demonstrating a dual advantage through its capacity to recycle waste materials and efficiently generate energy from them. Producing oxygen and handling carbon dioxide is another potential application of algal fuel cells which makes them ideal for integrating them in constructing habitats and colonies outside Earth. Therefore, developing innovative technologies like algal fuel cells can be pivotal in future space research.

## 9.2 ALGAL FUEL CELLS

Algal fuel cells are biochemical devices that have electrode compartments enclosed with a photosynthetic microorganism. Here, photosynthesis is carried out and they act as electron donors producing organic metabolites.[8] They can be single, dual, or three-chambered, built in a tubular, plate and frame, rectangular, circular, or H-type configuration made of Plexiglas or acrylic material to allow light penetration.[9] The algal fuel cell operates on the fundamental principle of substrate oxidation at the anode, leading to the generation of electrons, which then pass through a wire to undergo reduction at the cathode.[10]

### *9.2.1 CHAMBER CONFIGURATIONS*

Single-chamber MFCs are comparatively easy to operate and manage. In it, the anode chamber is separated from the air-cathode chamber by a gas diffusion layer (GDL) allowing a passive oxygen transfer to the cathode and eliminating the need for energy-intensive aeration of the liquid.[11] Air-cathode MFCs use free air as the electron acceptor.[12] It can also be modified to a membrane-less CCA.[13] Algae-powered MFC prepared with this configuration shows that it may offer the potential as a new technology to produce bioelectricity.[14] Then there are double-chambered MFCs with an anaerobic anode chamber and an aerobic cathode chamber, separated by a membrane. Bacteria at the anode called exoelectrogens oxidize organic matter and transfer electrons to the anode that pass through an external circuit producing current. Protons are exchanged through the membrane to the cathode combining with oxygen resulting in the production of water.[15] Three chambered MFCs are also there which consist of anode cathode and nourishment chambers.[16]

### 9.2.2 ELECTRODE CONFIGURATIONS

The metabolism of microbial particles results in the production of electrons which can be transferred directly or indirectly through an anode depending on the microbial catalysts used. Therefore, the design of the anode has greater importance in developing efficient MFCs and there have been huge improvements in electrode development over the last decade.[17] The electrode materials should have some general characteristics like their base materials must generally be of good conduction, good chemical stability, high mechanical strength, and low cost.[18] The most commonly used materials are carbon-based. They show very remarkable properties and good efficiency which makes them suitable for electrode purposes.[19] Carbon nanotubes are the best among them where CNT composites enhanced MFCs show increased performance than others.[20] Graphene hybrids with transition metal oxides are also used as electrodes since they show high performance and adhesion of biomaterial.[21]

### 9.2.3 PROTON EXCHANGE MEMBRANES

A proton exchange membrane (PEM) separates the anodic and cathodic compartments in a typical MFC. Proton exchange membranes affect a MFC's internal resistance and concentration polarization loss and in turn influence power output of the MFC.[22] The ratio of PEM surface area to system volume is important for the power output if it is below a critical threshold. The MFC's internal resistance decreases with the increase of PEM surface area over a relatively large range.[23] Nafion is a widely used PEM.

## 9.3 ALGAE IN SPACE RESEARCH

The first experiment that took algae to space was conducted by the Soviet Union in the Soviet Korabl-Sputnik 2 spacecraft in 1960. This indicates that research on algae-based applications in space operations commenced as early as the launch of the fifth satellite. The outcomes of this experiment were highly promising, revealing that algae exhibit the capability to execute all essential physiological processes even within the constraints of the space environment. Following this initial research, both the United States and Russia conducted comparable experiments to study the properties of algae in space, yielding identical results. Through the culmination of more than 50 experiments conducted over time by humanity, encompassing the transportation of algae into space, it has been established that fundamental physiological processes of algae, such as photosynthesis, growth, development, and reproduction, are indeed achievable in the space environment. Nonetheless, variations in the rates of these processes may occur due to extreme conditions.[24] It is seen that some phototrophs can even survive direct exposure to space conditions and among them, algal groups show very high tolerance.[25] Also in spaceflight conditions in the cabin atmosphere, optimum growth conditions can be found for algae (*C. vulgaris*) and certain modifications could accommodate a photobioreactor into the space cabin.[26] Studies show that integrating a photobioreactor could contribute to the oxygen and food supply of astronauts remarkably.[27] However, to incorporate a reliable photobioreactor, developments have to be made including the remote operations on bioreactor control and online monitoring.[28]

## 9.4 FUEL CELLS IN SPACE

Hydrogen-oxygen fuel cells are currently employed in space missions for energy production. Given that space vehicles employ hydrogen and oxygen in combustion processes, these fuel cells represent an augmentative asset, providing increased power and energy efficiency at the same mass.[29] Apart from this bio fuel cells are renewable sources of energy which is related to the life support system (LSS). It is seen that the incorporation of integrated fuel cells and photobioreactors as a hybrid system in LSS could benefit the production of energy, wastewater treatment, and production of food and oxygen.[30] However, microbial fuel cells have not often been into space. The first experiment that brought a bacterial fuel cell to space was conducted in the DELTA missions that utilized the Rhodoferax ferrireducens strain of bacteria to produce electricity. It proved the feasibility of a microbial fuel cell operating in space.[31]

Algal fuel cells stand out as highly beneficial microbial fuel cells, characterized by superior efficiency. They incorporate algae with hybrid capabilities, which could contribute to $CO_2$ and wastewater treatment. Algae show rapid growth and also can perform very well in unfavorable conditions. Algae-assisted microbial fuel cells have shown appreciable performance and net energy recovery of 11.5318 kWh $m^{-3}$ in outdoor conditions even when constructed with low-cost materials like rock phosphate-blended clayware and low-density polyethylene bags.[32] Another factor influencing the performance of algae is light. Photosynthetic microbial fuel cells with algae were able to generate electricity even in the absence of light showing stability as far as electricity production is concerned.[33] These features indicate that algal fuel cells qualify as a suitable candidate to experiment with electricity generation in space.

## 9.5 FACTORS IMPROVING THE PERFORMANCE OF AN AFC

### *9.5.1 ELECTRODE MODIFICATIONS*

Low efficiency is one of the major problems of algal fuel cells which kept them from practical upscaling in times past. There have been several studies going on to make better algal fuel cells of higher efficiency. The main factors that affect the performance of an AFC are temperature, pH, and the nature of electrodes. Apart from electrodes, we could find an optimal value for other factors and the enhancement offered by them is limited. The one thing that could solve this issue of reduced efficiency is developing more capable electrodes. An ideal electrode material should have high electrical conductivity, durability, stability, and a large surface area.[34] The adhesion of biofilms on electrodes has a major role in efficiency. Modifying the electrodes to increase the affinity of biofilms to have better adhesion of biomaterial to the electrode results in enhancement of efficiency.[35]

Treating many carbon electrodes with ammonia gas resulted in higher power density. Also, pretreatment with acid and heating techniques helps to generate cracks in electrodes and increases the surface area thereby delivering more power. Surface coatings are another extensive method used to improve the performance of electrodes. Apart from physical modifications coating the surface of electrodes can significantly increase the conductivity, thereby increasing the efficiency of the algal fuel cell.[36] Carbon-based electrodes modified in different ways show different efficiencies. However, they all exhibited a strong influence on system performance of an AFC.[37]

### 9.5.2 EFFECT OF LIGHT

We know that photosynthesis is exclusively a light-dependent process. Light has much impact on bio-reactions, thereby producing more electrons resulting in higher current and increased efficiency of the system. Different power outputs for the same system in different lighting conditions have been observed in studies. In one such study, a photosynthetic microbial fuel cell made with Spirulina platensis algae exhibited a relatively high power output under dark conditions[38] which contradicts the assumption that more light means more energy and shows dark cycle is advantageous too. A 16/8 light-to-dark cycle is optimal for increased performance of 12.7% than the intermittent 12/12 cycle rather than continuous illumination this is beneficial also.[39]

However, the light dependency of the system is not entirely based on the provided light itself. It is also related to the organic carbon substrates the algae consume for photosynthesis. A study shows that the effect of light on mixotrophic growth was more related to the substrate.[40] It is economical and energy-efficient to produce scale-up versions using direct sunlight as a light source.[41] This improves the scope of constructing algal fuel cells that work efficiently in space. Even in the absence of direct sunlight, space stations are constituted of efficient lamps even at the time of their construction,[42] right now fluorescent lamps are being used for lighting and solid-state LEDs are planned to replace them.[43] Any of these lighting systems are enough to produce an illuminated atmosphere for algal growth which shows that there is no need for additional lighting systems at space stations for algal fuel cells to perform efficiently. Finding the perfect lighting conditions and cycle for algal growth and photosynthesis can significantly improve the efficiency of an algal fuel cell. This can be achieved in space through direct sun exposure or external illumination.

### 9.5.3 USING DIFFERENT TYPES OF ALGAE

Distinct algae species exhibit varying rates of photosynthesis, consequently influencing the performance of algal fuel cells during their utilization. The widely used variety of algae for fuel cell purposes is *Chlorella vulgaris* which is a remarkable green eukaryotic microalga and was discovered as the first microalga with a well-defined nucleus.[44] Using different types of algae we can change the performance of the AFC. By finding the most efficient variety and configuration we could ideally create high-efficiency AFCs. Certain algal fuel cells with different algae and their performance are given below:

The change in power density is not exclusively because of the algae but the difference in all of the components. The same system used different algae[47] shows different output undoubtedly shows the importance of selection of algal variety for the construction of an AFC.

## 9.6 INTEGRATED HYBRID SYSTEMS

Hybrid life support systems can be considered an important part of the future interplanetary space research missions. Incorporating algae-based subsystems into this can benefit biotechnological as well as physico-chemical applications. These subsystems help bring closure of the carbon cycle in environment control and life support systems (ECLSS). The growth of algae has been studied under microgravity conditions to check the feasibility of this system, and very promising results have been obtained from it.[51] There have been studies going on for creating ideal life-supporting systems that can take human explorations way further into space. Many functional models of hybrid systems have been put forward in various studies. Water wall life support architecture is such a system incorporated with algae where the water walls approach provides a large number of repetitive, simple units to handle the same functions. Instead of continuously active large complex and expensive mechanical systems[52]

| Algae | Anode | Cathode | PEM | Power density | References |
|---|---|---|---|---|---|
| *Chlorella vulgaris* | Graphite plates | Graphite plates | Ultrex CMI 7000 | 34.2 mW $m^{-2}$ | [45] |
| *Spirulina platensis* | Graphite fiber felt | Graphite fiber felt | Sponge saturated with Zarrouk's media | 59.8 $mWm^{-2}$ (3 cells in parallel) | [46] |
| *Oscillatoria sp.* | Graphite plates | Graphite plates | Nafion 117 | 32.5 mW $m^{-2}$ | [47] |
| *Scenedesmus sp.* | Graphite plates | Graphite plates | Nafion 117 | 28.5 mW $m^{-2}$ | 47 |
| *Scenedismus obliqus* | Plain carbon paper | Pt coated Carbon paper | Nafion 117 | 153 mW $m^{-2}$ | [48] |
| *Scenedesmus quadricauda* | Carbon cloth with Ti | Carbon cloth with Ti | Glass wool | 62.93 mW $m^{-2}$ | [49] |
| *Scenedesmus/Chlorella vulgaris* (cathode) | Carbon cloth with Ti | Pt coated Carbon cloth | CEM CMI 7000 | 1926 mW $m^{-2}$ | [50] |

MeLiSSA created by the European Space Agency has a specific algae compartment that produces oxygen and supplements for the crew.[53]

Algal photobioreactors can be efficiently used for oxygen and food production on a future moon base. Algae is proven to be capable of withstanding continuous exposure to ionizing radiations[54] also apart from designing systems that work under microgravity there are no concerning difficulties for algae to perform its physiological functions. From studies, it is seen that about 30% of food consumption can be met by a 100 L photobioreactor for a single person that remarkably reduces the amount of food supplied from earth.[55]

Integrated systems for simultaneous treatment of carbon dioxide and bioelectricity production also can be considered a capable candidate for space explorations. Systems like bubbling-type photosynthetic algae microbial fuel cells provide an attractive strategy for $CO_2$ sequestration and bioenergy production. A maximum net energy of 1.824 kWh $m^{-3}$ was obtained in a B-PAMFC[56] which is a very considerable figure; however, it should be redesigned for space exploration for fail-proof establishments. The failure mode and cause studies show that the increase in the duration of deep space explorations requires a high level of system robustness. Most of the failures of these systems are caused by the cellular physiology of the algae, but fortunately these causes have well-established countermeasures.[57]

Algal fuel cells have been implemented in wastewater treatment plants, and considering the present-day space life support systems capable of recycling 90% of consumed water, the incorporation of an algae-assisted microbial fuel cell holds the potential to further enhance water treatment efficiency in such systems. They can be incorporated into any of these systems without any complex engineering but with slight redesigns. It is an additional value to the hybrid system by producing electricity from it.

## 9.7 CONCLUSION

Algal fuel cells are very promising as an emerging sustainable technology that would be very advantageous for space explorations. Introducing algae to space can be beneficial in many different areas including energy production and life support systems. Studies make sure the fact that algae can withstand extreme conditions and make it possible to grow in space.

There has been remarkable progress in the field of algal fuel cell research, but reduced efficiency and formidable logistics were prominent obstacles that kept it from practical experimenting in space. However, as research forwards developing and optimizing efficiency and implementing modern designs could bring solutions to this problem. The development and fabrication of high-performance electrodes for increased efficiency has been a rapidly growing field of study. By unlocking the complete capabilities of algal fuel cells, the mankinds significantly move closer toward building a sustainable outer earth settlement and being the only known inter-planetary species.

## KEYWORDS


- **algae**
- **algal fuel cells**
- **space research**
- **hybrid life support systems**
- **interplanetary colonization**
- **sustainable energy**


## REFERENCES


1. Liu, Z. et al. Carbon Monitor, a Near-Real-Time Daily Dataset of Global $CO_2$ Emission from Fossil Fuel and Cement Production. *Sci. Data* **2020,** *7* (1), 392. https://doi.org/10.1038/s41597-020-00708-7
2. Verduci, R.; Romano, V.; Brunetti, G.; Yaghoobi Ni, N.; Di Carlo, A.; D'Angelo, G.; Ciminelli, C. Solar Energy in Space Applications: Review and Technology Perspectives. https://doi.org/10.1002/aenm.202200125
3. Shukla, M.; Kumar, S. Algal Growth in Photosynthetic Algal Microbial Fuel Cell and Its Subsequent Utilization for Biofuels. http://dx.doi.org/10.1016/j.rser.2017.09.067
4. Gnaneswar Gude, V. Wastewater Treatment in Microbial Fuel Cells—An Overview Department of Civil & Environmental Engineering Mississippi State University. https://doi.org/10.1016/j.jclepro.2016.02.022
5. Vecchi, V.; Barera, S.; Bassi, R.; Dall'Osto, L. Potential and Challenges of Improving Photosynthesis in Algae. *Plants* **2020,** *9* (1), 67. Doi: 10.3390/plants9010067

6. Cycil, L. M. et-al. Investigating the Growth of Algae Under Low Atmospheric Pressures for Potential Food and Oxygen Production on Mars. DOI: 10.3389/fmicb.2021.733244
7. Shareefdeen, Z.; Elkamel, A.; Bin Babar, Z. Recent Developments on the Performance of Algal Bioreactors for $CO_2$ Removal: Focusing on the Light Intensity and Photoperiods. https://doi.org/10.3390/biotech12010010
8. Sivakumar, P.; Ilango, K.; Praveena, N.; Sircar, A.; Balasubramanian, R.; Sakthisaravanan, A.; Kannan, R. Microalgal Biotechnology. *Algal Fuel Cell.* DOI: 10.5772/intechopen.71228 (Chapter5). http://dx.doi.org/10.5772/intechopen.74285
9. Arun, S.; Sinharoy, A.; Pakshirajan, K.; Piet, N. L. Lens Algae Based Microbial Fuel Cells for Wastewater Treatment and Recovery of Value-Added Products. https://doi.org/10.1016/j.rser.2020.110041
10. Kannan, N.; Donnellan, P. Algae-Assisted Microbial Fuel Cells: A Practical Overview. *Bioresour. Technol. Rep.* 2021. DOI: 10.1016/j.biteb.2021.100747
11. Zielke, E. A. Design of a Single Chamber Microbial Fuel Cell. *Microbial. Fuel Cell* **2005,** 35.
12. You, S.; Zhao, Q.; Zhang, J.; Liu, H.; Jiang, J.; Zhao, S. Increased Sustainable Electricity Generation in Up-Flow Air-Cathode Microbial Fuel Cells. **2008,** *23* (7), 1157–1160. DOI: 10.1016/j.bios.2007.10.010
13. Zhuang, L.; Zhou, S.; Wang, Y.; Liu, C.; Geng, S. Membrane-Less Cloth Cathode Assembly (CCA) for Scalable Microbial Fuel Cells. **2009,** *24* (12), 3652–3656. DOI: 10.1016/j.bios.2009.05.032
14. Yuan, Y.; Chen, Q.; Zhou, S.; Zhuang, L.; Hu, P. Bioelectricity Generation and Microcystins Removal in a Blue-Green Algae Powered Microbial Fuel Cell. **2011,** *187* (1–3), 591–595. DOI: 10.1016/j.jhazmat.2011.01.042
15. Oh, S. E.; Min, B.; Logan, B. E. Cathode Performance as a Factor in Electricity Generation in Microbial Fuel Cells. **2004,** *38* (18), 4900–4904. DOI: 10.1021/es049422p
16. Gabryszewska, M.; Szewczyk, K. Properties of a Three-Chamber Microbial Fuel Cell. *Przemysl Chemiczny* **2011,** *90*, 192–197.
17. Kalathil, S.; Patil, S. A. Pant, D. Microbial Fuel Cells: Electrode Materials
18. Wei, J.; Liang, P.; Huang, X. Recent Progress in Electrodes for Microbial Fuel Cells. **2011,** *102* (20), 9335–9344. DOI: 10.1016/j.biortech.2011.07.019
19. Li, S.; Cheng, C.; Thomas, A. Carbon-Based Microbial-Fuel-Cell Electrodes: From Conductive Supports to Active Catalysts. *Adv. Mater.* **2017,** *29* (8), 1602547. DOI: 10.1002/adma.201602547
20. Kalathil, S.; Nguyen, V. H.; Shim, J.-J.; Khan, M. M.; Lee, J.; Cho, M. H. Enhanced Performance of a Microbial Fuel Cell Using CNT/$MnO_2$ Nanocomposite as a Bioanode Material. *J. Nanosci. Nanotechnol.* **2013,** *13* (11), 7712–7716. DOI: 10.1166/jnn.2013.7832
21. Zhao, C.-e; Wang, W.-J.; Sun, D.; Wang, X.; Zhang, J.-R.; Zhu, J.-J. Nanostructured Graphene/$TiO_2$ Hybrids as High-Performance Anodes for Microbial Fuel Cells. *Chem.: Eur. J.* **2014,** *20* (23), 7091–7097. DOI:10.1002/chem.201400272
22. Noori, M. T. Progress and Recent Trends in Microbial Fuel Cells Low-Cost Solutions for Fabrication of Microbial Fuel Cells: Ceramic Separator and Electrode Modifications, 2018; pp 95–124. DOI: 10.1016/B978-0-444-64017-8.00007-5

23. Du, Z.; Li, H.; Gu, T. A State of the Art Review on Microbial Fuel Cells: A Promising Technology for Wastewater Treatment and Bioenergy. **2007,** *25* (5), 464–482. DOI: 10.1016/j.biotechadv.2007.05.004
24. Niederwieser, T.; Kociolek, P.; Klaus, D. A Review of Algal Research in Space. *Acta Astronautica* **2018,** *146*, 359–367. DOI: 10.1016/j.actaastro.2018.03.026
25. Cockell, C. S.; Rettberg, P.; Rabbow, E.; Olsson-Francis, K. Exposure of Phototrophs to 548 Days in Low Earth Orbit: Microbial Selection Pressures in Outer Space and on Early Earth. **2011,** *5* (10), 1671–1682. DOI: 10.1038/ismej.2011.46
26. Niederwieser, T.; Kociolek, P.; Klaus, D. Spacecraft Cabin Environment Effects on the Growth and Behavior of Chlorella Vulgaris for Life Support Applications. *Life Sci. Space Res.* **2017,** S2214552417300780. DOI: 10.1016/j.lssr.2017.10.002
27. Svetoslav, A. Algal Research in Space: History, Current Status, and Future Prospects. *Innovare J. Life Sci.* **2016,** *4* (4), 1–4.
28. Fahrion, J.; Mastroleo, F.; Dussap, C.-G.; Leys, N. Use of Photobioreactors in Regenerative Life Support Systems for Human Space Exploration. *Front. Microbiol.* **2021,** *12*, 699525. DOI: 10.3389/fmicb.2021.699525
29. Burke, K. [American Institute of Aeronautics and Astronautics 1st International Energy Conversion Engineering Conference (IECEC)—Portsmouth, Virginia]—Fuel Cells for Space Science Applications, 2003. DOI: 10.2514/6.2003-5938
30. Belz, S.; Ganzer, B.; Messerschmid, E.; Friedrich, K. A.; Schmid-Staiger, U. Hybrid Life Support Systems with Integrated Fuel Cells and Photobioreactors for a Lunar Base. *Aerospace Sci. Technol.* **2013,** *24* (1), 169–176. DOI: 10.1016/j.ast.2011.11.004
31. de Vet, S. J.; Rutgers, R. From Waste to Energy: First Experimental Bacterial Fuel Cells Onboard the International Space Station; Z-Tec Publishing. *Bremen Microgravity Sci. Technol.* **2007,** *XIX*–5/6.
32. Khandelwal, A.; Chhabra, M.; Yadav, P. Performance Evaluation of Algae Assisted Microbial Fuel Cell Under Outdoor Conditions. *Bioresour. Technol.* **2020,** *310*, 123418. DOI: 10.1016/j.biortech.2020.123418
33. Gonzalez del Campo, A.; Perez, J. F.; Cañizares, P.; Rodrigo, M. A.; Fernandez, F. J.; Lobato, J. Characterization of Light/Dark Cycle and Long-Term Performance Test in a Photosynthetic Microbial Fuel Cell. *Fuel* **2015,** *140*, 209–216. DOI: 10.1016/j.fuel.2014.09.087
34. Zhou, M.; Chi, M.; Luo, J.; He, H.; Jin, T. An Overview of Electrode Materials in Microbial Fuel Cells. **2011,** *196* (10), 4427–4435. doi:10.1016/j.jpowsour.2011.01.012
35. Yang, Y.; Ye, D.; Liao, Q.; Zhang, P.; Zhu, X.; Li, J.; Fu, Q. Enhanced Biofilm Distribution and Cell Performance of Microfluidic Microbial Fuel Cells with Multiple Anolyte Inlets. *Biosens. Bioelectron.* **2016,** *79*, 406–410. DOI: 10.1016/j.bios.2015.12.067
36. Nitisoravut, R.; Thanh, C. N. D.; Regmi, R. Microbial Fuel Cells: Advances in Electrode Modifications for Improvement of System Performance. *Int. J. Green Energy* **2017**. doi:10.1080/15435075.2017.1326049
37. Scott, K.; Rimbu, G. A.; Katuri, K. P.; Prasad, K. K.; Head, I. M. Application of Modified Carbon Anodes in Microbial Fuel Cells. **2007,** *85* (5), 481–488. DOI: 10.1205/psep07018
38. Fu, C.-C.; Hung, T.-C.; Wu, W.-T.; Wen, T.-C.; Su, C.-H. Current and Voltage Responses in Instant Photosynthetic Microbial Cells with Spirulina Platensis. **2010,** *52* (2–3), 175–180. DOI: 10.1016/j.bej.2010.08.004

39. Bazdar, E.; Roshandel, R.; Yaghmaei, S.; Mahdi Mardanpour, M. The Effect of Different Light Intensities and Light/Dark Regimes on the Performance of Photosynthetic Microalgae Microbial Fuel Cell. *Bioresour. Technol.* **2018,** S096085241830539X. DOI: 10.1016/j.biortech.2018.04.026
40. Patel, A. K.; Joun, J. M.; Hong, M. E.; Sim, S. J. Effect of Light Conditions on Mixotrophic Cultivation of Green Microalgae. *Bioresour. Technol.* **2019,** S09608524 19303712. DOI: 10.1016/j.biortech.2019.03.024
41. Gouveia, L.; Neves, C.; Sebastião, D.; Nobre, B. P.; Matos, C. T. Effect of Light on the Production of Bioelectricity and Added-Value Microalgae Biomass in a Photosynthetic Alga Microbial Fuel Cell. *Bioresour. Technol.* **2014,** *154*, 171–177. DOI: 10.1016/j.biortech.2013.12.049
42. Lewin, I. Lighting in Outer Space: A Review of the Research and Development of the Exterior Luminaires and Lamps for the International Space Station. *Light. Res. Technol.* **2004,** *36* (3), 169–182. DOI: 10.1191/1365782804li111oa
43. Brainard, G. C.; Coyle, W.; Ayers, M.; Kemp, J.; Warfield, B.; Maida, J.; Bowen, C.; Bernecker, C.; Lockley, S. W.; Hanifin, J. P. Solid-State Lighting for the International Space Station: Tests of Visual Performance and Melatonin Regulation. *Acta Astronautica* **2013,** *92* (1), 21–28. DOI: 10.1016/j.actaastro.2012.04.019
44. Safi, C.; Zebib, B.; Merah, O.; Pontalier, P.-Y.; Vaca-Garcia, C. Morphology, Composition, Production, Processing and Applications of *Chlorella vulgaris*: A Review. *Renew. Sustain. Energy Rev.* **2014,** *35*, 265–278. DOI: 10.1016/j.rser.2014.04.007
45. Commault, A. S.; Laczka, O.; Siboni, N.; Tamburic, B.; Crosswell, J. R.; Seymour, J. R.; Ralph, P. J. Electricity and Biomass Production in a Bacteria- Chlorella Based Microbial Fuel Cell Treating Wastewater. *J. Power Sources* **2017,** S0378775317304032. DOI: 10.1016/j.jpowsour.2017.03.097
46. Longtin, N.; Oliveira, D.; Mahadevan, A.; Gejji, V.; Gomes, C.; Fernando, S. Analysis of *Spirulina platensis* Microalgal Fuel Cell. *J. Power Sources* **2021**. DOI: 10.1016/j.jpowsour.2020.229290
47. Naina Mohamed, S.; Jayabalan, T.; Muthukumar, K. Simultaneous Bioenergy Generation and Carbon Dioxide Sequestration from Food Wastewater Using Algae Microbial Fuel Cell. *Energy Sources, Part A: Recovery, Utilization Environ. Effects* **2019,** 1–9. DOI: 10.1080/15567036.2019.1666932
48. Kakarla, R.; Min, B. Photoautotrophic Microalgae *Scenedesmus obliquus* Attached on a Cathode as Oxygen Producers for Microbial Fuel Cell (MFC) Operation. *Int. J. Hydrogen Energy* **2014,** *39* (19), 10275–10283. DOI: 10.1016/j.ijhydene.2014.04.158
49. Yang, Z.; Pei, H.; Hou, Q.; Jiang, L.; Zhang, L.; Nie, C. Algal Biofilm-Assisted Microbial Fuel Cell to Enhance Domestic Wastewater Treatment: Nutrient, Organics Removal and Bioenergy Production. *Chem. Eng. J.* **2018,** *332*, 277–285. DOI: 10.1016/j.cej.2017.09.096
50. Cui, Y.; Rashid, N.; Hu, N.; Rehman, M. S. U.; Han, J.-I. Electricity Generation and Microalgae Cultivation in Microbial Fuel Cell Using Microalgae-Enriched Anode and Bio-Cathode. *Energy Conv. Manage.* **2014,** *79*, 674–680. DOI: 10.1016/j.enconman. z2013.12.032
51. Harald, H.; Jochen, K.; Gisela, D.; Stefan, B.; Reinhold, E.; Stefanos, F.; Heyer Arnd, G. High Density Long-Term Cultivation of *Chlorella vulgaris* SAG 211-12 in a Novel Microgravity-Capable Membrane Raceway Photobioreactor for Future

Bioregenerative Life Support in SPACE. *Life Sci. Space Res.* **2019,** S22145524 19300148. doi:10.1016/j.lssr.2019.08.001

52. Cohen, M. M.; Flynn Michael, T.; Matossian Renee, L. [American Institute of Aeronautics and Astronautics 43rd International Conference on Environmental Systems-Vail, CO] 43rd International Conference on Environmental Systems—Water Walls Life Support Architecture. DOI: 10.2514/6.2013-3517
53. Lasseur, C.; Brunet, J.; de Weever, H.; Dixon M. et al. MELiSSA the European Project of Closed Life Support System. *Gravit. Space Biol.* **2010,** *23* (2), 23–32.
54. Giuseppina, R.; Dania, E.; Mario, D.; Agnese, S.; Andrea, M.; Cecilia, F. et al. Ionizing Radiation Impacts Photochemical Quantum Yield and Oxygen Evolution Activity of Photosystem II in Photosynthetic Microorganisms. *Int. J. Radiat. Biol.* **2008,** *84* (11), 867–877. DOI: 10.1080/09553000802460149
55. Detrell, G. *Chlorella vulgaris* Photobioreactor for Oxygen and Food Production on a Moon Base—Potential and Challenges. *Front. Astron. Space Sci.* **2021,** *8*, 700579. DOI: 10.3389/fspas.2021.700579
56. Ming, L.; Minghua, Z.; Jianmei, L.; Chaolin, T.; Xiaoyu, T.; Pei, S.; Tingyue, G. Carbon Dioxide Sequestration Accompanied by Bioenergy Generation Using a Bubbling-Type Photosynthetic Algae Microbial Fuel Cell. *Bioresour. Technol.* **2019,** *280*, 95–103. DOI: 10.1016/j.biortech.2019.02.038
57. Matula, E. E.; Nabity James, A. Failure Modes Causes and Effects of Algal Photobioreactors Used to Control a Spacecraft Environment. *Life Sci. Space Res.* **2019,** *20*, 35–52. DOI: 10.1016/j.lssr.2018.12.001

# CHAPTER 10

# Bioenergy: Harnessing Nature's Potential for a Sustainable Future

REKHA MIRLE, PRIYANKA GHOW, KEERTHANA M., SWARUPA V., and ROOPASHREE R.

*Department of Chemistry and Biochemistry, School of Sciences, Jain University, JC Road, Bangalore, India*

## ABSTRACT

The term "bioenergy" refers to energy produced or generated from biomass, which comprises recently alive but now deceased organisms, primarily plants, wood, food crops like maize, energy crops, garbage from farms, yards, and woods are all examples of biomass that are frequently used for bioenergy. The carbon is liberated after combustion when this biomass is utilized to generate electricity and is discharged back into the atmosphere. Modern bioenergy is a fuel with almost zero emissions because as more biomass is produced, an equivalent quantity of carbon is absorbed.

## 10.1 INTRODUCTION

Bioenergy is one of the several choices available to help fulfil our energy requirements. Biomass, a sustainable energy source, is derived from recently deceased organic matter and is used for various purposes like the production of commodities, generation of heat energy, generation of electricity, and transportation fuels.[1]

Polymers and Functional Materials for a Cleaner Environment and Human Health: Sustainable Technologies and New Research. Ruby Varghese, Ann Rose Abraham, Shrikaant Kulkarni, & A. K. Haghi (Eds.)

Combustion of biomass results in the release of carbon, which subsequently re-enters the environment. Contemporary bioenergy is nearly completely free of emissions due to the increased absorption of carbon during biomass generation. With a share of 55% in the total renewable energy sector and contributing over 6% to the global energy supply, it stands as the largest renewable energy source on a global scale. Biomass has the potential to significantly contribute to the sustainable supply of energy needed in the future. At the moment, it is the leading supplier of renewable energy on a global scale, with significant capacity for expansion in generating heat, electricity, and transportation fuels.[2]

Unlike other renewable energy sources, bioenergy is distinct due to its involvement in the terrestrial carbon (C) cycle. Although emissions and sequestration may not always occur simultaneously, the carbon dioxide ($CO_2$) produced from bioenergy usage was previously stored in the atmosphere and will eventually be restored, provided that the bioenergy system is sustained. Terrestrial carbon stocks can be altered by the utilization of bioenergy, for instance, when a forest is converted into agricultural land to supply raw materials for biofuels.[3]

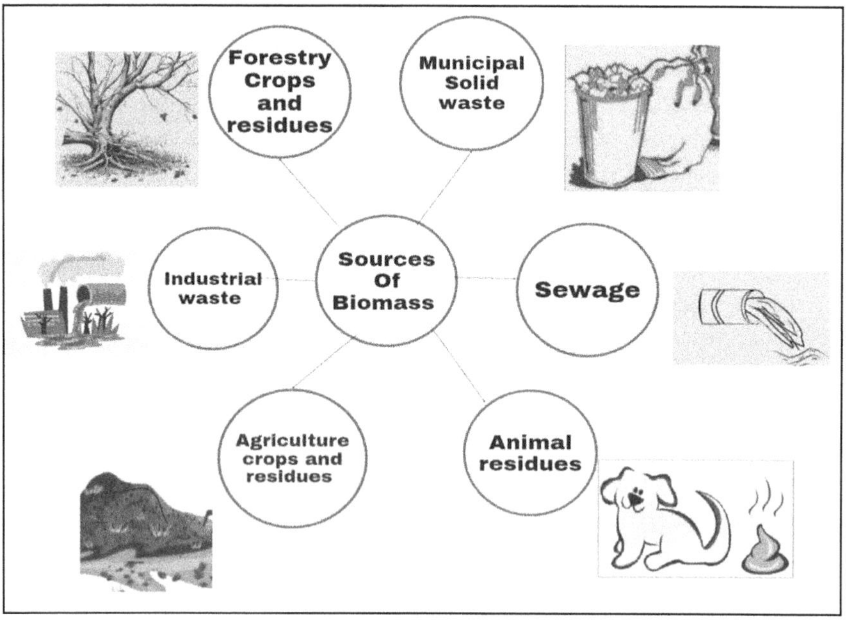


**FIGURE 10.1** Sources of biomass.

There are two types of bioenergy. Over 33% of the global population predominantly relies on it as an energy source, by burning wood fuel in cookstoves that are usually inefficient and produce pollution, as well as using dung and agricultural waste. Concurrently, it is the most rapidly expanding modern renewable energy source, generating power on a large scale to meet current energy needs and provide fuel for transportation. The combination of traditional and modern elements in bioenergy presents significant challenges to achieving sustainable development. The use of traditional bioenergy presents a significant problem due to its strong integration into the daily lives of impoverished populations in developing countries. While it plays a crucial role in providing essential energy services, it comes with substantial human, social, and environmental drawbacks. The transition from conventional biomass to cleaner and less harmful energy sources at higher levels of the "energy ladder" is long overdue. Modern bioenergy presents a challenge as it has the potential to become a significant source of global energy. However, there is no assurance that it will do this in an environmentally sustainable and socially desirable way. While bioenergy does include several potential benefits, it is also accompanied with significant risks. It is crucial to prevent the ironic adage "today's solutions are tomorrow's problems" from being applicable to bioenergy.[4]

## 10.2 VERSATILE CONVERSION TECHNOLOGIES

Various conversion pathways are employed to transform unprocessed biomass into practical forms of energy and to offer energy services such as heating, electricity, or transportation. Certain types of unprocessed biomass can be transformed into an intermediary biofuel prior to being utilized for a specific purpose. Various technologies, such as engines, boilers, refineries, turbines, fuel cells, and others, can be used to transform biomass and biofuels into energy. Biomass can be converted by thermochemical or biochemical processes, as well as chemical and physical methods as shown in Figure 10.2.

Biodiesel is a type of fuel for engines that is derived from rapeseed or other vegetable oils. It is manufactured through the process of methanol and other alcohol esterification. Biodiesel shares comparable characteristics with the conventional diesel fuel that is derived from fossil fuels. Physical techniques are commonly used to decrease the size of solid biomass, such as chipping. These processes can also be used to combine and compact microscopic particles into pellets or briquettes. Combustion is a very prevalent technique that liberates heat and has the capability to create electricity through the utilization of boilers and steam turbines.

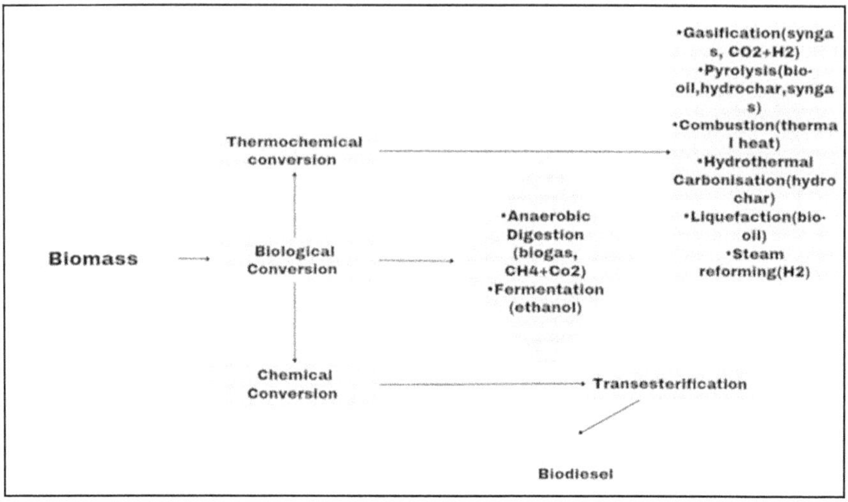


**FIGURE 10.2** Diverse conversion technologies.

Gasification procedures also provide thermal energy and electricity; however, they may need more stringent requirements for the raw materials used. They exhibit superior electrical efficiency and are particularly well-suited for installations above 10 MW, where there is a considerable demand for electrical energy and it may be sold at premium pricing. Pyrolysis refers to the anaerobic degradation of biomass at high temperatures, resulting in the production of solid, liquid, or gaseous compounds.

The composition and proportion of products, namely oil to char, are contingent upon the velocity and temperature of the procedure. Fast pyrolysis has the potential to generate bio-oil with an efficiency of up to 80%, whereas slow pyrolysis tends to result in a higher production of char. The byproducts of pyrolysis often consist of biofuels that may be conveniently stored or transferred for later use in different energy generation processes. Additional compounds/chemicals can be produced from pyrolysis oil, hence enhancing the economic viability of the process. Multiple biological processes are employed to generate gaseous and liquid fuels. Anaerobic digestion of sewage and effluents yields biogas, typically composed of methane and carbon dioxide. The unprocessed sedimentary residue might be dehydrated and incinerated or utilized as a soil enhancer. Additionally, several landfills generate biogas, and the existing designs enable the efficient capture and use of the released gas. Biomass sugars can undergo fermentation to yield bioethanol. Typically, this occurs after a hydrolysis pretreatment of the original biomass. The main sources of global bioethanol production are grain and sugar crops; however, there is currently a strong emphasis on researching and developing the use of woody biomass, known as lignocellulosic, for this purpose.[5,6]

## 10.3 ENERGY FOR REDUCED HOMES

As a percentage of the overall primary energy demand globally, biomass has remained essentially constant, moving from 12 to 11% between 1971 and 2004 even as the amount consumed in absolute terms increased by about 80%. Biomass continues to make up a sizeable portion of the primary energy supply in many developing nations—nearly half in Africa and more than 80% in several others (such as Nepal in Asia, Tanzania, Mozambique, and Nigeria in Africa), and if nothing changes in the world, it is anticipated that the number of families utilizing traditional biomass will continue to rise.

Although the modernization and general development of the energy industries have had some positive effects on their disadvantaged communities, there is often still more work to be done. Biomass continues to play an important, if not central, role in the human and economic fabric of emerging nations, especially in the rural and peri urban areas where a significant portion of their populations still reside. In fact, energy supply from biomass has increased in many nations virtually as quickly as or, in some cases, even faster than the energy sectors of developing countries over the previous few decades.[4]

## 10.4 LONG -TERM PROSPECTS FOR LAND-INTENSIVE BIOENERGY

Bioenergy contributes significantly to the energy economy, making up about 70% of the current supply of renewable energy and 9.5% of all primary energy.

The conventional utilization of biomass provides more than half of this bioenergy, predominantly for cooking and heating in homes, but also in small businesses (including ones such kilns for brick and charcoal). Despite this, there is still more to be done to make conventional biomass use more sustainable, effective, and safe for the welfare of individuals.

Modern bioenergy, also referred to as "bioenergy," contributed four times as much to the total amount of renewable energy used in 2017 than solar photovoltaic (PV) and wind energy put together. The majority of bioenergy is used to heat buildings and industries, but it is also anticipated that by 2023, bioenergy would produce 3% of the world's electricity and account for about 4% of the world's demand for transportation energy (International Energy Agency, 2018). Between 2010 and 2016, the growth pace of liquid biofuels to feed transportation was over 10% annually; nevertheless, the rate of economic expansion decreased to 4% annually throughout that period. The capacity of bioenergy electricity grew at an average annual rate of 6.5% between 2010 and 2016. It is anticipated that bioenergy (including liquid biofuels) will contribute 30% to the rise in the production of renewable energy between 2018 and 2023.

Most current scenarios for addressing climate change include bioenergy significantly. The Intergovernmental Panel on Climate Change (IPCC) in its latest Report on Global Warming observed that the biomass increased from 10% in 2020 to a median 26% of primary energy (154 EJ/year) in 2050 (with a range of 10 to 54%). By contrast, in 2050, a median of 22% of primary energy will come from solar and wind power.

A significant increase in bioenergy with carbon capture and storage (CCS) and the ongoing usage of high levels of bioenergy without it to the end of the century are predicted by many of the modeled paths included in that study. A 1.5°C future devoid of BECCS and with reduced bioenergy is predicted by some of the modeled pathways, but these rely on substantial behavioral changes (such as dietary adjustments), rapid technological advancement, a small global population, and a 32% reduction in overall energy demand by 2050 compared to levels in 2010. The positive framing that suggests large areas of land could be available for the development of bioenergy results in the important role that bioenergy and specifically BECCS plays in these scenarios.

A comparatively inexpensive and low-emission energy source, biomass avoids the intermittent nature of renewable energy sources like solar and wind power. Additionally, the market competition and the technological advancements make it doubtful that land-intensive bioenergy will continue to be a commercially viable energy source after 2050, even if all bioenergy were carbon neutral.[7]

## 10.5 BIOENERGY IN THE CURRENT CONTEXT

Even though modern and conventional bioenergy currently account for 9.5% of the world's primary energy supply, in 2100 they might only make up a small percentage of the energy mix. Both the amount and proportion of bioenergy are expected to increase dramatically. Demand for forest biomass will increase during the ensuing decades. The volume of wood pellets generated for biomass energy increased to 26 million tons (MT) between 2006 and 2015. In the EU, the primary importer of wood pellets, solid biomass accounts for nearly half (44.7%) of all renewable energy. East Asian markets for the new biomass are also growing quickly and may soon catch up to European demand.

Biomass use is growing due to regulations treating it as a zero-carbon fuel, leading to increased use of bioenergy. But over a ten-year period, only a percentage of the available biomass can assist the climate. With long carbon payback periods and near-term emissions higher than most fossil fuels, this might lead to a "double climate problem" and potentially reduce trees' capacity to fix carbon.

Global biofuel output is expected to increase to 142 million tons of oil equivalent (MTOE) by 2040 from 82 million MTOE in 2017. By 2030, there may be a further 18.6 MT of demand for palm oil in Indonesia, which increased its biofuel mandate to 30% in 2020. On the other hand, this might promote deforestation and raise transportation-related carbon emissions. There are special issues in managing the possible near-term expansion in bioenergy use while predicting a longer-term decline. Three categories of biomass supply exist: residues from other activities, removal for carbon storage or habitat improvement, and land-intensive bioenergy. Each category has different desirable, sustainability, and prospects, making it crucial to consider the potential for onsite and offsite carbon storage in different ecosystems.[7]

## 10.6 ADVANTAGES OF BIOENERGY

1. **It ensures a reliable and stable supply of electricity:** The United States imports fossil fuels at a daily pace of 7.86 million barrels, accounting for approximately 25% of global oil consumption. Utilizing biofuels for transportation lessens the 'country's reliance on foreign oil, as bioenergy is a domestically sourced resource. This not only reduces fuel expenses but also enhances our energy resilience by mitigating the risk of oil supply disruption due to natural disasters or political instability.
2. **It mitigates the release of greenhouse gases:** Bioenergy has the capacity to surpass all other available renewable energy sources in terms of reducing greenhouse gas emissions. The carbon dioxide emissions resulting from biomass combustion are balanced by the carbon dioxide uptake from plants and subsequent sequestration in the soil. It is biodegradable. Due to its origin from organic materials such as plant waste, biomass is inherently biodegradable, making bioenergy naturally degradable. Contrary to petroleum-based fuels, which are significant sources of pollution for the atmosphere, soil, and water, bioenergy is an environmentally friendly resource that may be generated indefinitely.
3. **It provides assistance to the agricultural sector:** Bioenergy not only lowers the energy expenses associated with agriculture, but also offers a valuable application for the organic waste generated from plants and animals on farms. Waste may be converted into biomass, which farmers can use as a source of income or to meet their personal energy needs.
4. **It stimulates employment opportunities:** Bioenergy production fosters employment opportunities in industry, agriculture, and various other sectors, now yielding around $50 billion in income and supporting nearly 300,000 jobs. Bioenergy will inevitably create more opportunities due to the 'industry's growth and the continuous progress of technology. The US economy is projected to gain approximately 1.1 million employments in the bioenergy sector by 2030.
5. **Biogenic energy offers significant advantages for agricultural producers:** Utilizing bioenergy for farming energy sources is a realistic and economical choice for farmers, as biomass is already

a natural byproduct of agriculture. In addition, farmers who consistently generate plant and animal waste have the opportunity to generate income by selling it to other sectors as biomass. FDCE has diversified its business operations since 2006 to include the production of bioenergy and biofuels. Discover 'FDCE's involvement in the bioenergy industry and reach out to us for further information on how you can benefit from this sustainable and cost-effective energy source.[8]

## 10.7 DISADVANTAGES OF BIOENERGY

1. **Expensive:** The generation of bioenergy can be accompanied by a substantial cost. Initially, the establishment of bioenergy facilities necessitates a substantial initial capital outlay. Additionally, one must take into account the expenses associated with the collection and transportation of biomass materials. In addition, the organic material must be stored upon arrival to the facility, necessitating additional financial resources. As a result, biomass is more costly compared to other available renewable alternatives. Unlike other renewable energies such as solar, these frequent transit and storage expenditures are not necessary. After the construction of a solar farm, there is no requirement for fuel transportation or the storage of sunshine. Despite the higher cost of biomass compared to other forms of renewable energy, it remains more affordable than fossil fuels. Harvesting biomass incurs far lower costs compared to the extraction and mining of fossil fuels.
2. **Space requirement:** Plants that produce bioenergy take up a lot of space. mostly due to the necessity of accommodating storage facilities. This imposes restrictions on the potential locations for constructing biomass energy facilities. Certain biomass plants cultivate their own organic matter. These plants require ample room to cultivate crops or establish miniature woodlands. They also require a larger land area to generate each kilowatt hour of electricity.
3. **Greenhouse gas emissions:** Bioenergy has no carbon footprint. However, carbon dioxide emissions from biomass persist in the atmosphere. Carbon-zero renewable sources are highly advantageous in addressing climate change. Reducing the emission of

carbon dioxide is highly desirable. Not only carbon, but also other substances are emitted into the atmosphere via bioenergy. Nitrogen oxides, carbon monoxide, and methane are the main greenhouse gases released during the combustion of wood and other biomass fuels. Because methane gas retains approximately thirty times more heat than carbon dioxide, it plays a major role in both air pollution and climate change.

4. **Environmental impact:** Although bioenergy is a renewable resource, it is not the most environmentally friendly alternative available. Dependence on biomass energy has the potential to result in catastrophic deforestation. The increased need for resources by biomass plants may lead to accelerated deforestation of forests, additionally, hence increasing the vulnerability of the affected areas to erosion. Some biomass plants rely on algae, which frequently require large volumes of fertilizers including phosphorus in order to produce the required levels for energy production. The excessive utilization of fertilizers can result in significant harm to nearby water bodies and animals, particularly avian species. The utilization of plants for biomass energy may lead to an augmentation in the application of fertilizers and pesticides, hence exerting adverse effects on the surrounding ecology.
5. **Inefficient:** Currently, bioenergy exhibits lower efficiency compared to alternative energy sources. Often, the energy required to incinerate the organic matter exceeds the energy output of the process. In order to utilize biomass on a wide scale, it is imperative that new technologies are developed to enhance its efficiency.[9]

## 10.8 BIOENERGY BY 2050

The percentage of the theoretical potential (i.e., the greatest biomass that might theoretically be achieved under just biophysical constraints) that is achievable with existing technology is referred to as the "technical bioenergy potential" from here on. Due to the lack of a standardized approach, estimations of the technological bioenergy potential vary. Assuming a "food/fiber first principle" and excluding deforestation, after accounting for a variety of environmental constraints, the bulk of recent research evaluating technical bioenergy potentials eventually produce an estimate of the "environmentally sustainable bioenergy potential."

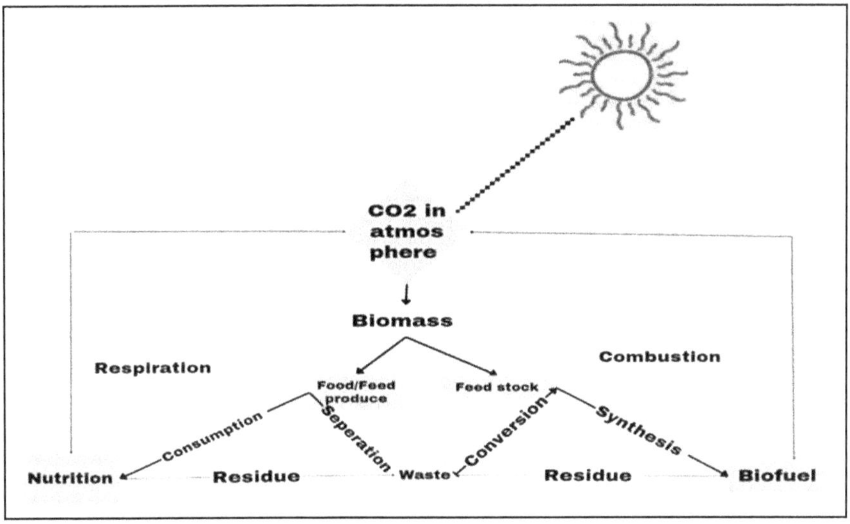


**FIGURE 10.3** Features of biomass energy technology.

Recent estimates range from 50 EJ yr1 to >1000 EJ yr1, almost three orders of magnitude, depending on this revised definition of the world technical primary biomass potentials in 2050. For example, the Global Energy Assessment estimated that global technological bioenergy potentials might reach 50–500 EJ yr1 in 2050, while the SRREN indicated a range of 160–270 EJ yr1. Although there has 'not been agreement on the size of the future worldwide technological bioenergy potential as a result of the discussion after the publishing of these global assessments, it has helped in gaining further insight into a few of its numerous structural aspects.[10]

## 10.9 BIOENERGY AND POLICY OBJECTIVES

Bioenergy possesses significant potential to significantly augment its present role in attaining policy objectives, such as mitigating $CO_2$ emissions and guaranteeing energy stability, while simultaneously contributing to social and economic advancement aims. In order to identify the most efficient strategies for reducing greenhouse gas (GHG) emissions using bioenergy, it is critical to comprehend the emissions produced by various bioenergy methods and to recognize the role that bioenergy plays in reducing emissions within particular businesses. Various bioenergy pathways can display substantial disparities in greenhouse gas (GHG) emissions.

Substituting biomass for fossil fuels in heat and power production is generally more economically efficient and leads to greater reductions in emissions per unit of biomass compared to using biomass instead of gasoline or diesel for transportation. However, the bioenergy industry that does not move or change may rely on various low carbon alternatives. Biofuels continue to be the primary option for reducing carbon emissions in road transportation until electric and/or hydrogen fuel cell cars become widely used. This is not anticipated to happen for many decades.

Biofuels are likely to be the only feasible option for mitigating carbon emissions in the aviation sector in the future, given the difficulty of finding an alternative to liquid fuels in this industry. Land suitable for growing biomass for energy production can also be used to create biospheric carbon sinks. The comparative attractiveness of these two options is influenced by various aspects, particularly land productivity, which encompasses co-products, and the efficiency of substituting fossil fuels. Furthermore, a detailed assessment of the potential direct and indirect emissions resulting from the conversion of land to new uses is required. These emissions have the potential to significantly impair the environmental benefits of bioenergy and carbon sink projects. Transportation will not transition to electric and/or hydrogen fuel cell cars on a large scale for many decades.

The appraisal of carbon reduction potential is significantly influenced by the time frame. The sink option is generally more advantageous on a shorter time scale, whereas a longer time scale enables larger savings. This is due to the fact that biomass production is not constrained by saturation and may consistently achieve reductions in greenhouse gas emissions by substituting fossil fuels, from one harvest to another. It is feasible to effectively oversee mature forests that have ceased to actively absorb carbon dioxide in a conventional manner for the purpose of producing wood and other forest-related goods, leading to a somewhat modest decrease in greenhouse gas emissions per unit area. Alternatively, they may be converted into energy-intensive crops with higher productivity (or utilized for food cultivation), but this would require the release of at least a portion of the created carbon reservoir.

The use of native biomass resources can help ensure energy stability, depending on the particular energy source it is replacing. The act of importing biomass from different foreign origins aids in the diversification of the energy composition. However, changes in biomass production and mismatches between supply and demand in the food and forest product industries can have an impact on supply stability, potentially leading to shortages. The production of bioenergy can result in many environmental and socioeconomic outcomes, which can be either advantageous or disadvantageous. The production of biomass feedstock is primarily responsible for the majority of environmental impacts. However, these impacts can be mitigated by implementing optimal technologies and appropriate laws. Technical solutions can be employed to mitigate the environmental impacts of bioenergy conversion facilities. Ensuring the presence and efficient enforcement of sufficient environmental legislation is the crucial element in implementing these solutions.

By harnessing organic waste, agricultural/forestry remnants, and lignocellulosic crops that may be grown on many soil types, we can reduce the demand for land and water resources and mitigate competition with food production. Feedstock production systems can also offer multiple benefits. Harvesting forest waste improves the conditions of the forest site for replanting. Thinning generally promotes the growth and productivity of the remaining, while removing biomass from densely populated areas reduces the risk of wildfires.

Harvest helps reduce greenhouse gas emissions by substituting fossil fuels. It is feasible to employ conventional methods to utilize mature forests, which have ceased to absorb carbon dioxide, for the production of timber and other forest commodities. Nevertheless, this strategy yields a minimal decrease in greenhouse gas emissions per unit of land. Alternatively, they can be converted into plantations that require a significant amount of energy and produce higher yields, or utilized for food production. Nevertheless, this would necessitate the liberation of at least a fraction of the carbon reservoir that has been amassed.

The anticipated integration of electric and/or hydrogen fuel cell vehicles into the transportation industry is not projected to happen for several decades. In the realm of agriculture, biomass can be cultivated in versatile plantations. By employing meticulous site selection, design, management, and system integration, these plantations offer supplementary environmental advantages. These advantages, in return, increase the worth of the systems. The development of bioenergy policy should be synchronized with the attainment of environmental and social objectives.

Regulating bioenergy is essential to effectively handle environmental and social concerns, recognize and value the environmental advantages of bioenergy systems, and actively contribute to rural development objectives.[6]

## 10.10 CONCLUSION

In the global energy landscape, biomass and bioenergy are essential because they provide sustainable and renewable alternatives to fossil fuels. Biomass, which is derived from organic resources such as wood, agricultural wastes, and organic waste, is a flexible resource that may be used to produce biofuels, heat, and power. Technological and application advances in the use of biomass and bioenergy are promising for a bright future in a number of sectors. Biomass offers a sustainable solution to combat global warming by generating heat, power, and biofuels through various conversion processes like gasification, combustion, and advanced technologies like anaerobic digestion and pyrolysis. The future of bioenergy lies not only in the production of electricity but also in fuels for transportation. Fuel blends already include biofuels like ethanol and biodiesel, which help to reduce reliance on fossil fuels and promote cleaner transportation networks. Furthermore, the creation of sophisticated biofuels from non-food sources like garbage and algae has great potential for a sustainable energy future.

## KEYWORDS

- **bioenergy**
- **biomass**
- **sustainability**
- **advanced technologies**
- **carbon**
- **environment**
- **renewable energy**

## REFERENCES


1. https://www.energy.gov/eere/bioenergy/bioenergy-basics
2. https://www.iea.org/energy-system/renewables/bioenergy
3. Bioenergy, I. E. A. Bioenergy, Land Use Change and Climate Change Mitigation. *Report for Policy Advisors and Policy Makers. IEA Bioenergy: ExCo, 3*. https://www.researchgate.net/profile/Goeran-Berndes/publication/280076283_Bioenergy_Land_Use_Change_and_Climate_Change_Mitigation_Background_Technical_Report/links/55a678a208ae92aac77f29d3/Bioenergy-Land-Use-Change-and-Climate-Change-Mitigation-Background-Technical-Report.pdf
4. Sagar, A. D.; Kartha, S. Bioenergy and Sustainable Development. *Annu. Rev. Environ. Resour*. **2007**. https://doi.org/10.1146/annurev.energy.32.062706.132042
5. Bioenergy, I. E. A. Benefits of Bioenergy. *IEA Bioenergy: ExCo*. 2005. https://www.ieabioenergy.com/wp-content/uploads/2013/10/52_BenefitsofBioenergy.pdf
6. Bioenergy, I. E. A. Bioenergy—A Sustainable and Reliable Energy Source; International Energy Agency Bioenergy, Paris, France, 2009. https://agroparistech.hal.science/hal-01590366/
7. Reid, W. V.; Ali, M. K.; Field, C. B. Development. *Annu. Rev. Environ. Resour.* **2009**. https://doi.org/10.1146/annurev.energy.32.062706.132042
8. The Future of Bioenergy. Global Change Biology. https://doi.org/10.1111/gcb.14883.
9. https://fdcenterprises.com/what-are-the-benefits-of-bioenergy/
10. https://www.solarreviews.com/blog/biomass-energy-pros-and-cons
11. Creutzig, F. et al. Bioenergy and Climate Change Mitigation: An Assessment. *Glob. Change Biol*. **2014**. https://doi.org/10.1111/gcbb.12205

# Index

## O



## P

For Product Safety Concerns and Information please contact our EU
representative GPSR@taylorandfrancis.com
Taylor & Francis Verlag GmbH, Kaufingerstraße 24, 80331 München, Germany

www.ingramcontent.com/pod-product-compliance
Ingram Content Group UK Ltd.
Pitfield, Milton Keynes, MK11 3LW, UK
UKHW022204190926
475702UK00003B/203
* 9 7 8 1 7 7 9 6 4 3 6 5 0 *